LAND AND WATER TRANSPORTATION

General Editor: Anne O'Daly

Published by Brown Bear Books Ltd
4877 N. Circulo Bujia, Tucson, AZ 85718
USA
and
Studio G14, Regent Studios, 1 Thane Villas,
London N7 7PH, UK

Cataloging-in-Publication data available upon request.

ISBN 978-1-83572-062-2 (ALB)
ISBN 978-1-83572-068-4 (paperback)
ISBN 978-1-83572-074-5 (ebook)

Children's Publisher: Anne O'Daly
Design Manager: Keith Davis
Picture Manager: Sophie Mortimer
Consultant: Chris Woodford

Manufactured in the United States of America
CPSIA compliance information: Batch#AG/5664

Picture Credits

STEM

Key Concepts in STEM: Engineering and Technology describes the scientific principles and feats of engineering that have shaped the world we live in. Since earliest times, people have created tools and machines to provide shelter, warmth, prosperity, security, and protection from disease. The history of transportation dates back thousands of years to when people first started to domesticate animals. Innovations rapidly followed, with sailing boats, steam power, and the internal-combustion engine. This book describes the technological advances that made progress possible. Explanatory diagrams and informative photographs explain how STEM (science, technology, engineering, and math) has transformed our lives and continues to influence our world today.

Contents

ANIMAL POWER

Since the dawn of civilization humans have harnessed the strength of large animals to help them move around and carry heavy loads. Horses and other animals are still used today where road and rail would be impossible.

No one knows when people first began to use tame animals to carry things, but it was the first cities were probably being built around 6,000 years ago. It may have been in the settlements that grew up on the fertile plains of the Middle East, where the terrain is well-suited to animal transport.

Animal-powered vehicles are not in common use today in most parts of the world. However, they are sometimes used for ceremonies and leisure activities such as tourist sightseeing.

Lands Without the Wheel

The wheel was probably invented several times over in different parts of the world. However, the ancient civilizations of the Americas, such as the Maya, Aztecs, and Incas, never used wheels. The likely reason for this is that wheeled wagons need strong draft animals such as oxen, good roads, and flat land. American animals, such as llamas, are too small to haul large carts. Instead, these animals carry loads through mountains that are too rugged for wheeled vehicles to cross. In snowbound countries, too, wheels were useless: reindeer were used to pull sleds in Scandinavia in the winter. Even in the Middle East, where the wheel was invented, camels were used to make journeys across the roadless deserts.

Oxen were probably the first animals to be used, then donkeys. Oxen were dragging wooden sleds in Mesopotamia (now part of modern Iraq) before 3500 BCE. A rope through a nose ring is all that is needed to lead a tame ox. They can also be harnessed simply: a pair of oxen can be placed on each side of a single shaft, with a wooden crosspiece resting as a yoke across their shoulders. In North Africa, donkeys are still ridden without any bridle. To guide it, the rider touches the donkey lightly between its ears with a stick.

The Wheel

Oxen helped people move heavy loads, but they were slow and the load was limited to what could be dragged along rough tracks. After 4000 BCE, tree trunks began to be used as rollers, making loads easier to pull. At around 3500 BCE, in Mesopotamia, wheels were invented. The first wheels were solid wooden disks, either from a single plank or from two or three planks fastened with cleats (wedges of wood or metal).

INVENTING THE WHEEL

Before wheels were invented, loads were moved on sleds that ran over log rollers. A sled rolling across a log many times will cut a groove, and this might be how the first wheels and axles were invented.

The narrow groove became the axle, while the wide roller formed the wheel. The axle was then attached to the sled, forming a simple cart. Later, carts had fixed axles in which only the wheels turned.

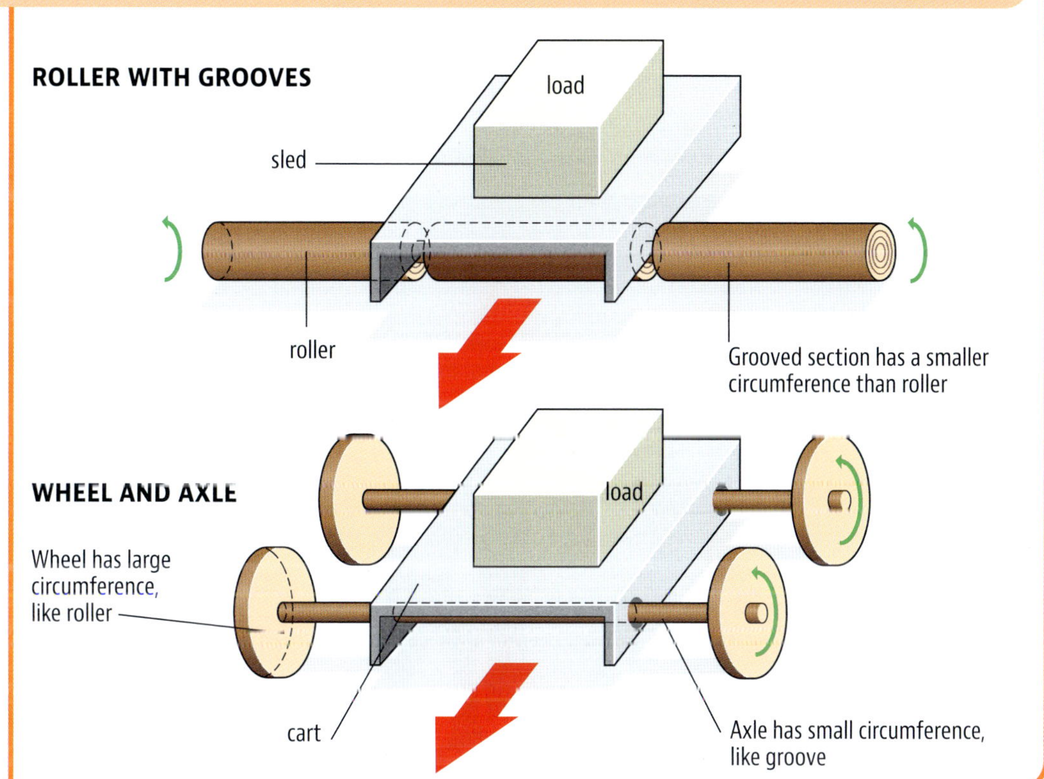

WAR CHARIOTS

Wheeled vehicles were soon being used in warfare. In Mesopotamia, both four-wheeled and two-wheeled carts were used as mobile platforms from which soldiers could hurl spears. Solid wooden wheels made them heavy and cumbersome, but by 1900 BCE. the Mesopotamians had developed the spoked wheel—a circular outer ring supported by wooden spokes radiating from its center. This resulted in light, fast-moving vehicles that could dart about the battlefield.

Horses evolved for life on the wide open grasslands, such as here in Mongolia. The first domestic horses were probably tamed in this Central Asian region.

Horses for Transportation

Almost as important as the wheel was the discovery of how to tame horses so they could be ridden. Riding probably began about 5,000 years ago with nomadic herders on the steppes (flat plains) of Central Asia.

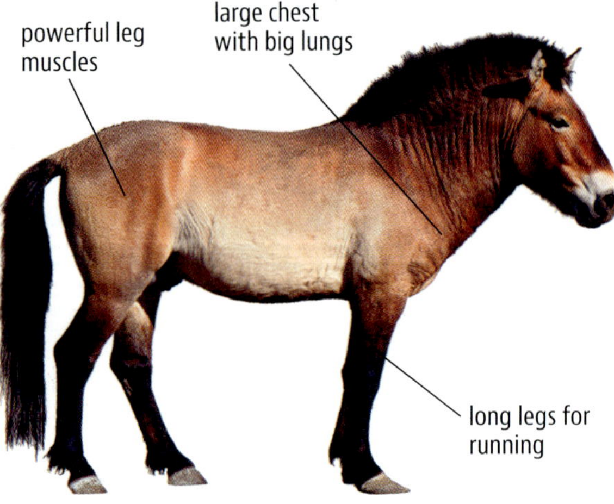

ANIMALS FOR TRANSPORTATION

Horses combine strength and mobility, which makes them the most versatile animals for human transportation. A number of other species are also important. In the Arctic Circle teams of dogs or reindeer are used by various peoples to pull sleds. Extreme heat and lack of water are lethal to horses, but camels thrive in such harsh environments and are vital to the desert nomads (roaming peoples) of Central Asia, North Africa, and the Middle East. A camel can carry a heavy load for 30 miles (50 km) in a single day, using only one sixth of the water that a horse would consume in a similar journey. For really heavy work, however, there is no better animal than the elephant. In India, elephants are used to push down trees and transport logs and other heavy loads. Elephants were once also ridden into battle, where they inspired terror in opposing troops, and opposing horses!

Mountainous or rocky terrain can prove difficult for horses to cross. People sought more sure-footed alternatives. The donkey was one, but its small size limits the loads that it can carry. However mules, produced by breeding a male donkey with a female horse, combines the useful qualities of both animals. The true mountain experts, however, are the llamas of South America and the yaks of the Himalayas, which have thick coats to protect them from the cold.

Camels have a store of fat in their humps, which allows them to survive without eating or drinking for several days.

Riding horses gave nomadic peoples far greater mobility, and they could make longer journeys and herd animals to better grazing lands far more easily. They could also tower over their enemies in battle, and make faster escapes.

These early riders probably rode bareback, gripping with their knees and using just a simple strap to control their mounts. Stirrups and bridles had not yet been invented. Riders also placed animal skins on their horses' backs to make the ride more comfortable—so creating the first basic saddles.

Horseshoes are heated so they can be bent and hammered into the right size and shape before being nailed to a hoof.

SADDLE TECHNOLOGY

In 2023, archaeologists found the remains of a leather saddle in a grave in northwestern China. The saddle had been stuffed with a mixture of straw, deer hair, and camel hair. Tests showed that the saddle had been made between 724 and 396 BCE.

Scythian saddles, made by people in Eurasia between about 724 and 396 BCE, had padded leather and felt and hoops of wood. The next development was the saddle tree, which spreads the weight of the rider over the horse's back. The Han Chinese in around 200 BCE and the Romans in around 100 CE probably used saddle trees. Modern Western saddles still have many of these early features.

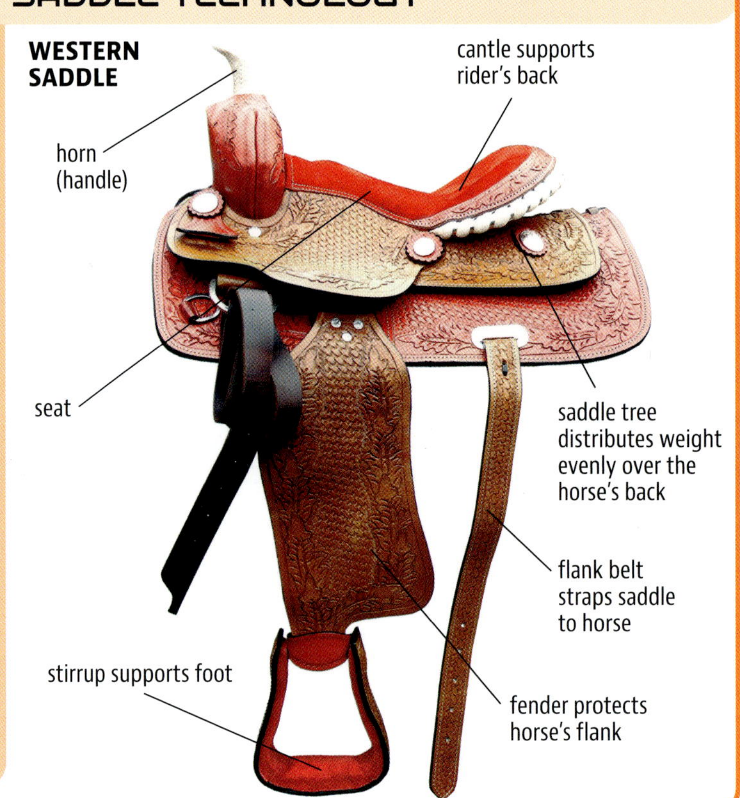

Charioteers

Early domestic horses were smaller than modern breeds. Before the invention of the saddle tree and the stirrup, it was difficult for a heavily armed soldier to fight from horseback. Horse-drawn chariots were a solution: they made use of a horse's mobility but were stable enough for the soldiers in them to fight and to fire weapons. The Hyksos people used a chariot-based army to invade Egypt around 1700 BCE. The Hittites from Asia Minor are said to have used 3,500 chariots to defeat the Egyptian forces of pharaoh Ramses II.

Around 1300 BCE the first simple metal bits were used (it is almost certain that wooden or bone bits were used before this, but none have been found). The bit is a piece of metal

FACTS AND FIGURES

• Modern horses are descended from the Dawn Horse, *Hyracotherium*, a tiny creature little more than 12 inches (30 cm) high that lived some 50 million years ago.

• One of the earliest known books was about horses. Written by Kikkuli, who lived in the Hittite Empire in 1400 BCE, it explained how to use horses to pull chariots.

• Horses were useful for transport but there was a downside. In New York, for example, there were 120,000 working horses around 1900. They produced 1,300 tons (1,200 tonnes) of dung per day, which had to be swept up and disposed of.

DOUBLE BRIDLE

Several types of bridle are used today. This shows the most complex, the double bridle, which is used for events such as dressage (horse training and maneuvering) competitions, where the rider needs extra control. Two separate bits go in the horse's mouth, each with its own set of reins. The snaffle bit, with rings at each end, goes in first. This is the bit that would be used on a single bridle as the main aid for controlling the horse. The curb bit is H-shaped, with sides that project up and down, and a chain and strap that go behind the mouth. When the curb rein is pulled, the center of the bit presses on the mouth, while at the same time the sides are pulled backward by the reins at the bottom, making the top (eye) tilt forward. The eye pulls on the top of the horse's head through the poll strap, and on the horse's jaw through the curb chain.

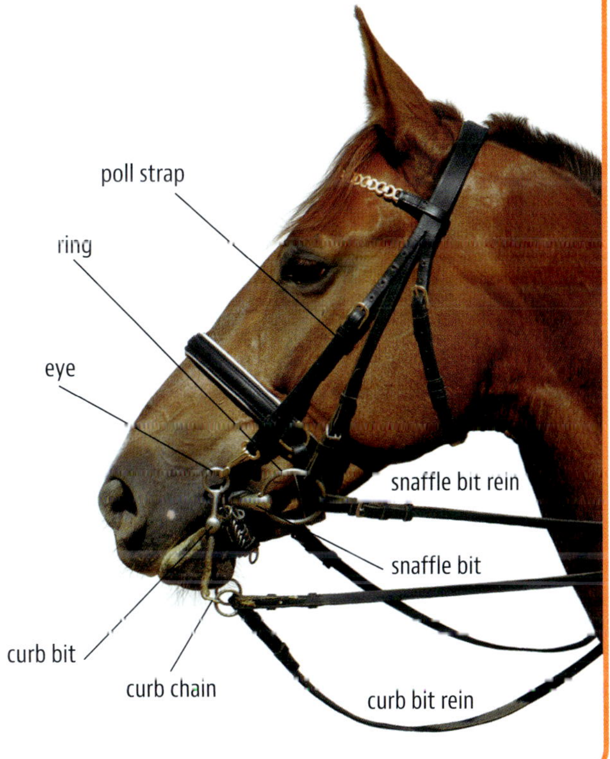

Collar harnesses put the weight of the load on the horse's shoulders rather than the neck or throat. A horse can pull four times the weight with a padded collar than it can with the earlier, more primitive, yoke harness.

placed between the horse's teeth that can be pulled using bridle rings and leather straps called reins to give commands to the horse.

Single bits remained the standard way of controlling horses until the fourth century BCE, when the Celtic tribes of northwest Europe introduced the curb bit. This is an H-shaped bit with a cross-piece in the horse's mouth. It pulls the horse's head down, giving greater control. With better control of the horse, chariots became less important in warfare and cavalry, soldiers on horseback, took over as a major way of fighting battles.

A New Age of Warfare

By the second century CE a nomadic Central Asian people called the Sarmatians were using saddles that had arches to fit over the horse's back and high peaks front and rear so the rider was held very securely. In warfare this allowed the horseman to use a heavy weapon, such as a lance or long sword.

THE COACH

According to popular belief it was wagonbuilders in the Hungarian village of Kocs who, during the 15th century, revived the idea of using small front wheels on wagons, allowing the vehicles to turn more easily and making them more stable by placing the center of gravity farther back. Soon they added the key refinement, suspending the body of the vehicle between the axles rather than resting it on them— so absorbing some of the shocks of the road. The *kocsi*, or coach, was born.

Coaches spread throughout Europe in the 16th and 17th centuries. At first these modern, comfortable vehicles were used only by the richest families. Later, larger coaches were used as long-distance public transportation.

STIRRUPS

Stirrups, light frames that support the rider's feet, were also invented in the second century CE. They were a huge step forward in the use of horses for riding. They seated the rider more securely on the horse and reduced the level of skill required to ride at top speeds. The consequences of saddles and stirrups were still clear to see 1,000 years later in the Middle Ages. A big horse fitted with a high-peaked saddle was the preferred mount for wealthier soldiers. However, merchants, pilgrims, priests, and others could also ride on smaller horses, their feet secure in their stirrups with their goods on pack mules.

Stirrups allow a rider to stand up while on horseback, making it a lot easier to stay balanced and jump obstacles.

COACH SUSPENSION: MASS AND INERTIA

The first passenger coaches were supported on leather slings, like a hammock. Then crescent-shaped metal springs called leaves were added, making a huge difference. Not only was the ride more comfortable, but the coach was much safer to drive at speed, floating over bumps rather than catapulting off them. Springs work because the heavy carriage has so much inertia that it takes a lot of force to move it. When the coach goes over a bump, the wheels and axles go up in the air. Most of this motion is absorbed by the bending springs, and very little moves the carriage body itself. How well this works depends on how bouncy the springs are and on the ratio of the sprung weight (the body) to the unsprung weight (wheels and axles).

THE ROADBUILDERS

At the end of the 18th century, roadbuilders devised new ways of making roads. Two Scottish engineers, Thomas Telford (1757–1834) and John MacAdam (1756–1836) led the way. Telford's technique was to dig a trench into which he put a foundation of heavy rock topped with 6 inches (15 cm) of compacted stones. MacAdam believed in lighter construction. He laid roads made of small stones (below). Workmen checked the size of these stones with their mouths—if they were too big to fit in the mouth, they were too big for the road! His surface was finished by adding tar—a mixture called tar-macadam, now shortened to "tarmac."

In Europe and North America the stagecoach was a common form of public transport in the 17th and 18th centuries. The horses or mules had to be changed frequently.

The Romans used chariots for racing but preferred infantry (foot soldiers) and cavalry for warfare. They invented an important piece of horse technology, the horseshoe—a U-shaped metal plate that is nailed onto the horse's hoof to prevent it from being worn down on hard surfaces, such as the Roman road network.

Carts and Wagons

For most of the Middle Ages, the carts of Europe were little different from Roman times. They usually had four wheels all the same size, no suspension, and no steering— even though steerable carts with pivoting front axles had been invented by the Celts in 50 BCE.

Since horses had been introduced to the Middle East and Europe, various harnesses had been used, but none was ideal. They forced the horse to pull with its neck rather than its body, and could easily ride up and press on its windpipe. The horse collar, a broad harness that rested over the horse's shoulders, was invented in China and allowed a wagon, plow, or canal boat to be pulled with maximum efficiency. Even so, wagons in 16th-century Europe struggled along at far less than walking pace. Not only were they crude constructions, but they had to run on terrible roads. The Roman roads had long since

deteriorated, leaving tracks that were bumpy and dusty in summer and muddy quagmires in winter. If they could afford it, most people still preferred to travel on horseback.

This changed with the invention of the coach, a comfortable design of cart that could be pulled at high speed—if the road surface allowed it. The use of coaches spread throughout Europe in the 16th and 17th centuries, although only the richest families could afford this new technology—and all the horses that were needed to pull them.

Wells Fargo and the Pony Express

Wells Fargo was set up in 1852 by Henry Wells and William G. Fargo to provide banking and transportation during the California Gold Rush. Its stagecoaches carried passengers, mail, and gold across the American West.

The Pony Express was a service that carried mail between St. Joseph, Missouri, and Sacramento, California—a route of 1900 miles (3058 km). Riders worked in relays, switching horses every 10 to 15 miles (16 to 24 km). Delivery took just 10 days. The Pony Express operated from April 1860 to October 1861. Wells Fargo became an agent for its mail service. However, the completion of the transcontinental telegraph in October 1861 provided a much quicker and more convenient form of communication, and the Pony Express services were redundant.

Road and Rail

Better roads gradually became essential during the 18th century. In France, 15,000 miles (24,000 km) of new roads were built using conscripted (forced) labor.

From the end of the 18th century roadbuilding proceeded rapidly all over Europe and the eastern part of North America. By 1830 there were 20,000 miles (32,000 km) of

HORSES AND RAILROADS

Innovation is not a sudden process. Old and new continue to exist side by side, and different uses are often found for the old. The first railroads, for example, did not replace the horse—for they used horse-drawn cars. Long before railroad engines had been invented, horses drew loads along rail tracks. Mostly they worked in mines, hauling coal or iron ore; but the first passenger railroads were also horse drawn. The very last horse-drawn railroad service, at Fintona in Northern Ireland, operated until 1957. When steam railroads did arrive at the beginning of the 19th century, they quickly became the biggest owners of horses. Horses were used for shunting rolling stock, unloading freight cars, and taking freight onward to its destination. As late as 1928, the London Midland and Scottish Railway in Britain owned over 9,600 horses.

good roads in Britain alone. A journey that once took eight days now took less than two. Mail coaches traveled at high speeds, and many types of private carriages were developed, designed for the wealthy.

In the 1840s, horse-drawn carriages were superbly constructed, with two- and four wheeled carriages transporting people in style. But the age of the railway had dawned, and horse-drawn transport was eventually replaced by trains.

HUMAN POWER

In a world of high-speed cars, supersonic aircraft, and space travel it's easy to forget that most of our journeys still rely on the oldest power source—our own bodies.

Before people drove, sailed, and flew, they walked, sometimes for days on end, covering hundreds of miles. It is likely that the first shoes were made 30,000 years ago during the Stone Age. These were simply hardened animal skins tied to the soles of the feet with

The mountain bike, sometimes shortened to MTB, was invented in the 1970s. It is used for cycling on rough off-road terrain.

RUBBER BOOTS

Waterproof rubber boots are known by a number of different names, from rainboots to gumboots. However, to many people they are known as Wellingtons, or "wellies." This name is derived from the Anglo-Irish soldier, Arthur Wellesley, the 1st Duke of Wellington (1769–1852). In the early 19th century he designed a calf-length cavalry boot, made from leather, which was waterproof but also easy to fit into stirrups. Wellington's boot was also the inspiration for the cowboy boot.

Rubber wellies were first manufactured in the 1850s.

strips of leather. In cold places the skin was pulled around the ankles for warmth. Shoes like this evolved into the moccasins that were popular during the Iron Age and that were more recently used by Native Americans.

Around the world people developed shoes for different purposes, making the best use of local materials. In the desert areas of the Middle East, leather sandals were used to protect feet without making them too hot;

The Japanese geta is half sandal, half clog. During wet weather, the high "teeth" keep the foot raised above any puddles.

in Europe and Japan wooden clogs were ideal footwear for working in muddy fields, while in China, a country with less access to leather, shoes were made from woven fabric.

On Your Bike

Many inventors came up with plans for human-powered machines, but it wasn't until 1817 that a design came along that would become the bicycle we know today. The German inventor Karl von Drais (1785–1851) realized that it was possible to balance a machine on just two wheels as long as it was moving forward. His wooden *Laufmaschine*

Some modern houseshoes, or slippers, still use the same design as traditional animal-skin moccasins. However, modern moccasins often have a hard rubber sole.

SLIDING ALONG

The first skis were made about 4,500 years ago in Norway and Sweden as a way of spreading the weight of a person so that they didn't sink into the snow. Skis soon developed into an effective means of transport. Early skis were covered with animal fur, arranged with the hairs pointing backward. As the ski moved forward the hair of the fur slid easily over the snow, but when skiers pushed back on the ski, the hairs dug in to give good grip, making it possible to walk uphill. Early skiers also developed the sledge. Initially just an animal skin dragged along the ground. Later, sledges were made with wooden or bone runners to reduce friction.

Modern skiing developed in the mid-19th century in Norway and then in the Alps, the mountain range that divides France, Switzerland, and Italy. This sport became popular with the invention of the chairlift in the 1930s. Today's skis are made from carbon fiber, which is light and strong. Wax is rubbed onto the skis to fine tune them for different snow conditions.

Downhill skis are bound tightly to the skier's stiff, plastic boots. This rigid connection allows the person to control the position of each ski at high speed. The boots also support the skier's ankle in the event of a crash.

A design of Karl von Drais's early pedal-free "running machine" from 1832.

(running machine) had no pedals but was pushed along by the rider's feet. On a downward slope the rider's legs rested on a bar, and the machine would coast along effortlessly. Von Drais's invention started a craze that swept through Europe and the United States. Riding schools were set up to teach enthusiasts—mainly young men— how to ride.

AN OLD IDEA MADE NEW?

Time and again when people look for the earliest examples of machines that we now take for granted, the same name crops up— the Italian artist Leonardo da Vinci (1452–1519). The bicycle could be no exception. Some experts think that the first design for a bicycle was among the drawings on a Leonardo manuscript from the 1490s. However, others believe that the drawing of a two-wheeled cycle with pedals and a chain is a fake. Another curiosity is a 16th-century stained-glass window in a church in Stoke Poges, England, which appears to show an angel riding on a two-wheeled machine.

Soon, improved bicycle designs began to appear. The best was probably the Pedestrian Curricle, made by Englishman Dennis Johnson in 1819. Better known by its nicknames, the "hobby horse" and "swift walker," Johnson's design used metal and was lighter than von Drais's original. Johnson also made a model for women, with a low frame to accommodate long dresses.

CYCLING RACES

The pedal-powered bikes, known as velocipedes, began a craze that has continued to the present day: cycling races. Riding schools opened to cater to the growing bands of new riders. The first race for velocipedes was held at Saint-Cloud Park, near Paris, on May 31, 1868, and in November of the following year a race was held over 83 miles (134 km) of public roads between Paris and Rouen. Both races were won by an Englishman called James Moore—the first successful racing cyclist. The first six-day bicycle race in America was held in Maddison Square Garden, New York City, in 1891 on a board track. Today's most important cycling race—the Tour de France—began in France in 1903.

Today's cycling road races may take place over many days and involve dozens of riders. The riders complete different stages of the race each day, so there are stage winners and finally an overall winner.

Pedal Power

In 1839, Scottish blacksmith Kirkpatrick MacMillan created a version of the hobby horse with pedals. These pedals were connected to cranks that turned the rear wheel when the pedals were pushed backward and forward. This machine was very clumsy and heavy. In the 1860s, a French carriage-maker called Pierre Lallement attached pedals and cranks to the front wheel of a hobby horse. At last here was a machine—named the velocipede ("speedy foot")—that could be efficiently powered by a human.

In the 1870s, Englishman James Starley (1830–1881), designed the Ordinary bicycle, better known as a "penny farthing." The rider sat high off the ground on a huge front wheel, steering with almost vertical forks. The bike frame was a tube that followed the curve of the front wheel, and it had a tiny rear wheel.

This pedal-powered velocipede from the 1870s had no chain (or brakes). Instead the pedals were connected directly to the front wheel by a crank.

PENNY FARTHINGS

Why was the front wheel on a penny farthing so big? The answer was speed. Cycling began as a sport, and everyone wanted to be quickest. In the days before gears were added, when pedals were attached directly to the driven wheel, the only way to make a faster bike was to fit it with a bigger wheel. With every turn a big wheel covered more ground than a small wheel. The only restriction on the size of wheel was the length of the rider's legs. The champion of the 1880s was Herbert Lidell Cortis—a man whose height of 6 ft 2½ inches (1.89 m) allowed him to race bikes with front wheels 60 inches (152 cm) high. Modern adult-sized bikes have wheels that are generally 26 inches (66 cm) in diameter. The back wheel was small to keep the weight down. Since the front wheel provided the drive and steering, the rear wheel was only needed for stability.

The Ordinary bicycle was nicknamed the "penny farthing" because its wheels looked like large and small coins of the day.

When a chain to drive the rear wheel was invented, bicycles were designed with two wheels of the same size.

Despite the popularity of the Ordinary bicycle, many people found it too difficult to ride. Short people could not reach the pedals, and women in their long skirts were unable to ride for fear of becoming caught in the spokes. Even the greatest supporters of the machine found it hard to stay balanced without falling off.

Many people tried to build safer bicycles. Chains or levers were used to gear-up a smaller front wheel so that one turn of the pedals made the wheel turn more than once. At first these machines were ridiculed by riders of Ordinarys. But eventually the popularity of these "safety bicycles" led to the demise of the Ordinary.

SOCIETY AND INVENTIONS

Transport of the Future

The bicycle has often seemed like a poor alternative to the car. Open to the weather, not able to hold much luggage, and usually unable to carry more than one person, bikes haven't always seemed as comfortable or as convenient as cars. This view, however, is changing. Many people are increasingly concerned about the amount of pollution and congestion cars create in our cities. With this concern, an increasing interest in healthy, more environmentally friendly modes of transportation has developed. This is where the bike begins to look like a better option. Small in size, making no pollution, and promoting health, bicycles are now taken seriously by many urban planners interested in improving the quality of life in cities. The result of this is an increase in dedicated "cycling-only" routes.

Special cycle routes have been built in many major cities to protect riders.

The machine that changed the shape of bicycles forever was the Rover Safety, introduced in 1885 by the nephew of James Starley, John Kemp Starley. The Rover Safety set the shape of bikes as we know them today. The key innovations of the Rover Safety were that the rear wheel was driven by a chain, and a strong frame replaced the single metal tube of the Ordinary. Other people made their own "safetys," adding refinements.

A bike chain runs from a cog wheel that is connected to the pedals to a series of gear cogs on the rear wheel.

DERAILLEUR GEARS

A bicycle's gears transmit rotary motion from the pedals to the back wheel via the bicycle chain. Derailleur gears were invented in 1911 and use an ingenious mechanism that lets a rider change gear (to get more speed or climbing force) without stopping or getting off. They consist of a series of cogs attached to the drive shaft, or axle, of the rear wheel and a tension pinion that moves the chain from one cog to another.

To achieve maximum efficiency, the cyclist tries to keep the effort and rate of pedaling constant. On a flat surface, a high gear is selected, and the tension pinion moves the chain to a small cog. One turn of the pedals now rotates the rear wheel several times, making the bike go faster. On a hill, however, the effort required to pedal in high gear is too great, so the cyclist selects a lower gear (larger cog). The bike now moves a smaller distance for each turn of the pedals, but the effort needed to turn the pedals is smaller.

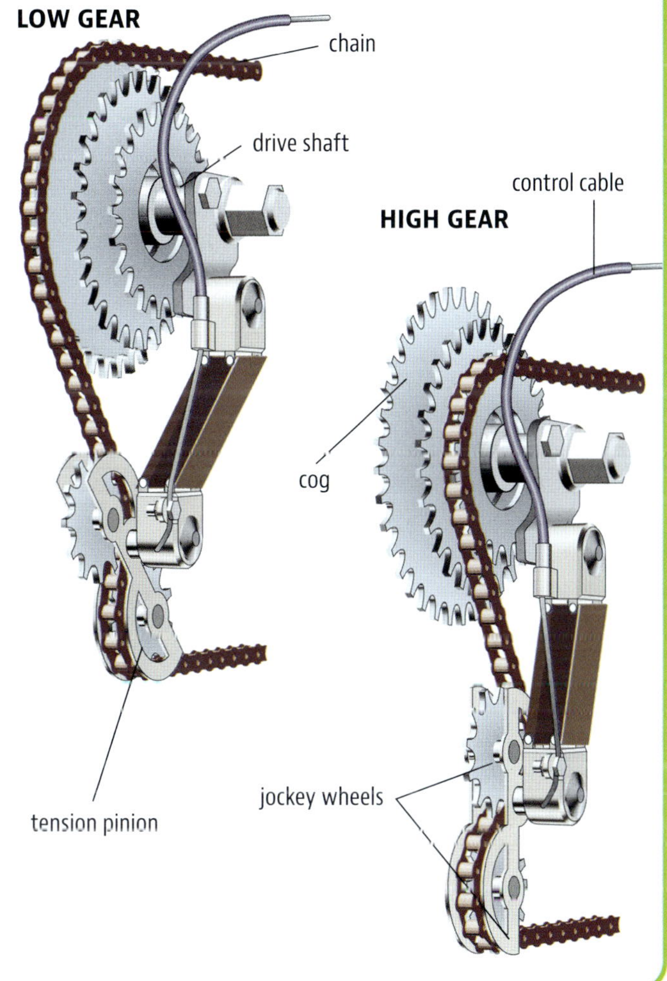

After some experimentation wheels became equal sized, the diamond-shaped frame was introduced, and finally the air-filled tires were added. These were invented in 1888 by John Boyd Dunlop (1840–1921) who fitted rubber tubes to his son's tricycle. By the 20th century, the safety was the standard bike design, and bicycle design has changed little since then.

Although the basic shape of bicycles stays the same, many variations have been tried. For example, in 1962 British car designer Alex

BMX stands for bicycle motocross. BMX bikes are small, light, and fast. They are used for aerial stunts and balancing tricks.

RECUMBENT BIKES

Although invented in the 1930s, a recumbent bicycle still turns heads on a modern street. In this design, the rider lies almost flat, peddling feet first. The recumbent bicycle is streamlined, with the rider traveling feet-first, and creates less drag than conventional bikes. The pedals are at the front of the machine, linked to the rear wheel by a long chain. The large seat and reclined riding position make this a comfortable bike, suitable for long journeys. However, the rider sits low in motor traffic, making it more difficult to see—and also more difficult to be seen.

SUPERLIGHT BIKES

Drag force is created when an object moves through a fluid—slowing it down. In world-class sprint cycling races, high drag can make the difference between winning and losing. Top cyclists sometimes spend many hours in a wind tunnel, perfecting their riding position. Today, all top sprint cyclists have superlight bikes, often with covered wheels, and teardrop-shaped crash helmets to reduce drag and make forward motion as easy as possible.

Moulton added a rubber suspension to a bicycle, and that made it possible to fit tiny 16-inch (40-cm) wheels highly geared to work as efficiently as the larger wheels of conventional machines. The rider sat high above the main frame, making it easy to get on and off, and it took up less storage space. Similar bikes have been built with folding lightweight frames, so they can be stowed inside a bag and then carried on to another mode of transport. The bicycle is a technology that has continued to develop since its invention 200 years ago. It is a relatively clean form of transport and it does not rely on fossil fuels–just human muscle power.

E-BIKES, ROAD, AND TRAIL BIKES

The basic design of the bicycle has been extensively modified to suit different uses and conditions. An electric bike (e bike) has an electric motor to provide extra power for the cyclist. A trail bike must be able to cope with rough terrain. It has a robust frame, with shock absorbers fitted to the front and rear forks for a smoother ride over bumps and potholes. Many use disk brakes that act at the center of the wheel to give control over steep inclines and slippery surfaces. Racing bikes are built for speed and should only be used on roads. Lightweight frames and small caliper brakes are used to save weight.

Road bikes have narrow, smooth tires that grip a flat surface while cutting drag, but are hard to steer on rough and muddy ground. Trail bikes have wide, gripped tires that work on all terrains, but do create more resistance on the bike.

TRAINS AND TRACKS

Rail travel has a long history. The first tracks were built 500 years ago for hauling rocks inside underground mines. Today, railroads criss-cross the world carrying passengers and freight.

The first passenger rail services were hauled by horses, but the large scale of long distance railroads demanded a mechanical source—steam engines. The first steam-powered vehicles to run on rails were built by Richard Trevithick (1771–1833), a British engineer. In February 1804, a locomotive he had designed pulled five wagons loaded with 11 tons (10 metric tons) of iron over 9.75 miles (14 km) of track at a speed of around 5 mph (8 km/h). The load also included some 70 spectators, who had climbed aboard to enjoy the unusual ride through southern Wales.

Trevithick's locomotive was primitive but pointed to the future. For example, it had

smooth wheels, which still gripped the rails to the surprise of many critics. It also had an

THE FIRST RAILROADS

Railroads existed long before steam trains. Mine tracks were invented in Europe around 1550 CE. Parallel rails were used to move coal or iron ore. Horses, ponies, or even children were used to pull the wagons. At first the tracks were just made from wood. Later, strips of iron were fixed to the rails. The first solid iron rails were made in England in 1738. In 1789 movable switches (points) were invented so that wagons could roll from one track to another track.

Tracks make it easier to pull loads through rough mine tunnels.

The modern railroad is the result of many linked inventions. Some developed the locomotives, some the engines that drive them, while others control how trains operate on a busy network.

efficient boiler that generated steam at high pressure. This was one of the breakthroughs that made the new age of rail transportation possible. Previously, steam engines had been too big and bulky and lacked power. Another problem was that these first steam engines were too heavy for the cast iron rails, which often broke, resulting in the train crashing. This was one reason why there was no immediate rush to build railroads. Four years after his initial demonstrations, Trevithick constructed a railroad to demonstrate the power of steam. His "Catch-Me-Who-Can" engine ran on a circular track, and anyone who dared could ride on it after paying a modest fee. But the show was a failure, and Trevithick went bankrupt.

STREETCARS

The streetcar was invented by English engineer John Outram in 1775. His "tram" ran on rails and was drawn by two horses. In 1832, John Stephenson built the first streetcar line, again horse-drawn, in New York. It closed after three years. The main problem was that early streetcars ran on rails raised from the road, which obstructed other traffic. But in 1852, French engineer Émile Loubat found a way of embedding the rails into the road surface and used this technique to build the Sixth Avenue Line in New York. Thirty years later, cities had become congested with horse-drawn traffic. Something cheaper and cleaner was needed. In 1888, the engineer Frank J. Sprague (1857–1934) built the first electric streetcar line in Richmond, Virginia. Within 10 years, 25,000 miles (40,000 km) of electric streetcar lines were in use in cities across the United States.

Yellow electric streetcars, some made from wood, have been running through Lisbon, Portugal, since 1901.

LOCOMOTIVE DESIGNS

Many of the decisions of George and Robert Stephenson had a lasting impact on railroad design, including the width of the gauge still used on most railroads today. Another innovation was the steam-blast design, which directed exhaust steam through a narrow pipe into the chimney. Air was pulled in after the steam, increasing the draught through the furnace. This made the coal in the furnace hotter, increasing power and speed.

The Stephensons show off their Rocket, *the locomotive that pulled the first passenger train service in 1829.*

Railroad Pioneers

The first steam-powered locomotives started running a regular railroad service on a short, narrow track between the English city of Leeds and the mining district of Middleton. The Middleton Railway had been built in 1758 to carry coal. It used horses until 1812, when a locomotive designed by English engineer John Blenkinsop (1783–1831) began to haul cargo—and occasionally passengers. Despite Trevithick's example, it ran on toothed rails,

FACTS AND FIGURES

• The standard-gauge railroad track of 4 ft 8½ in (1,435 mm) is the width of a Roman chariot's wheels. Today, standard gauge accounts for about 60 percent of railroad tracks worldwide.

• The widest track ever built was the 7 ft (2,134 mm) gauge developed by the British engineer Isambard Kingdom Brunel (1806–1859). He wanted extra width to give greater stability at the high speeds he was aiming for.

• Iron rails gave way to steel ones in the 19th century. In the 1950s, rails were welded together into long lengths. This gave a smoother ride.

• Completed in 1916 and covering 5,750 miles (9,250 km), the Trans-Siberian Railroad is the longest continuous railroad in the world.

• The total length of railroad routes reached its peak in 1917, with about a million miles (1,600,000 km) worldwide.

As well as passengers, England's Liverpool and Manchester Railway carried livestock and heavy cargo. The railroad connected the sea port of Liverpool with the factories of Manchester.

LAYING TRACK

As well as developing locomotives and rail cars, George Stephenson established the standards to which railroads were built. He believed that the railbed should be meticulously prepared,and that the inclines (grades) should be less than one percent, so early steam engines would be able to pull long, heavy trains. Curves should have a radius of half a mile (0.8 km) or more to prevent derailments. His ideas were followed worldwide, as foreign engineers visited England. However, his reliance on flat and straight tracks was unsuited to building railroads through rugged landscapes worldwide.

The Liverpool and Manchester Railway in England ran across Chat Moss Bog, bringing steam technology to an agricultural landscape unchanged for centuries.

not smooth ones. But it attracted the attention of George Stephenson (1781–1848), who was convinced that he could do better.

Beginning in 1814, Stephenson built a succession of locomotives, gradually improving the design. As a result, he was made construction engineer for the Stockton and Darlington Railway. Completed in 1825, this was the first specially built steam railroad that anyone could use for shipping freight, rather than being owned and used by a single mining company. It led to a much bigger project—a railroad linking the English cities of Manchester and Liverpool. Stephenson supervised the line's construction and in 1829 he won the competition for the best engine with his *Rocket* (in fact, largely designed by his son Robert, 1803–1859). This locomotive used a new and more powerful design of boiler and had a top speed of 16 mph (26 km/h) when pulling a train and 36 mph (58 km/h) on its own.

United States Railroads

The first American steam-powered passenger line, the Charleston & Hamburg Railroad, opened on December 25, 1830. It ran on a 6-mile (9.6-km) stretch of track, using a locomotive called the *Best Friend of Charleston*. In the United States, long distances meant that the railroads had to be cheap to build—the invention of the wooden crosstie and T-cross-section rail (requiring less metal than a solid rail) made this possible. Mountain crossings meant sharper curves and steeper grades. At the peak of its route over the Alleghenies, the Baltimore & Ohio Railroad climbed 1,800 ft (600 m) in a 17-mile (27 km) stretch, a grade of more than two percent.

Locomotives had to be redesigned to cope with these conditions—using the earlier British designs only resulted in frequent derailments. One early innovation was the use of a swiveling leading truck to guide the locomotive into the tighter curves. Passenger

A long freight train twists and turns through the Canadian Rockies.

STEAM LOCOMOTIVE

In a steam locomotive, fuel (wood, coal, or oil) is burned in a firebox directly in front of the cab. Pipes carry the hot air and smoke from the firebox through the boiler, a large tank holding water, to the smokestack. The water starts to boil, forming steam in the dome. From the dome high-pressure steam passes to alternate sides of the cylinders, driving the pistons. Connecting rods link the pistons to the driving wheels. A safety valve allows steam to escape if the pressure becomes too great. A tender at the back carries supplies of fuel and water.

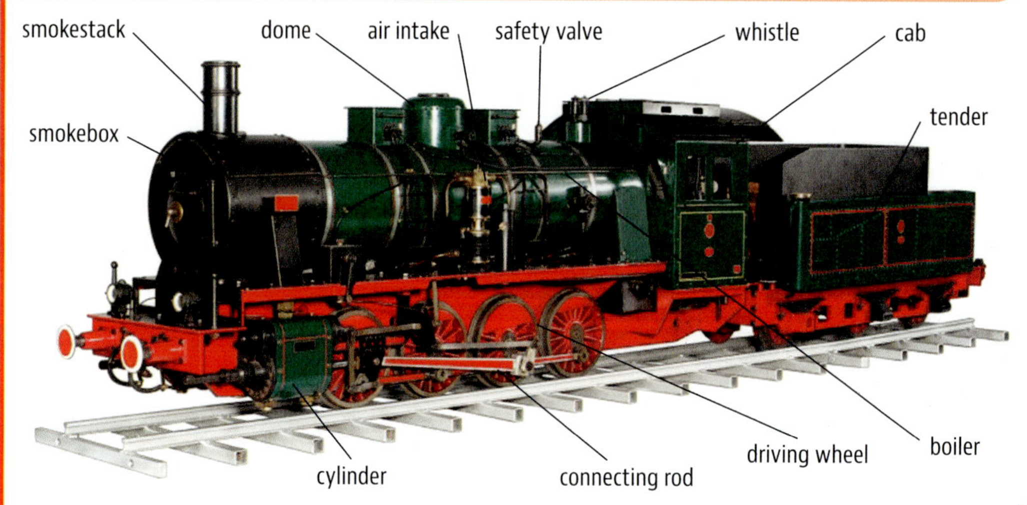

and freight cars, too, were soon built on four-wheel swiveling trucks—called bogies—rather than fixed axles. These and many other innovations produced railroads that were strong, cheap to operate, and simple to maintain.

American locomotives had a plow-shaped cowcatcher at the front to push animals off the track.

Comfort for Passengers

The first sleeping cars, called bed carriages, were installed on trains running from London to Lancashire, England, in 1838. Dining cars first ran in the United States in 1863 between Baltimore and Philadelphia. Two years later, the American industrialist George M. Pullman (1831–1897) patented his own design of sleeping car. In 1867 he founded his own company, the Pullman Palace Car Company, to construct luxury passenger accommodation. The scenic dome, or vista dome, car was a Russian invention, first seen in 1867.

Subways and Elevated Tracks

The first subway (underground railroad) was opened in London in 1863 to ease congestion on the city streets. Its steam trains transported ten million passengers in its first year. Elevated railroads get people off the sidewalks and can be built more easily than subway railroads. The first electric elevated city railroad was built in 1893 in Liverpool, England.

Replacing Steam Engines

The first diesel locomotive was built for Prussian State Railways in 1912. It proved a disappointment: the mechanical transmission

Subway trains carry many thousands of people every day.

SIGNALLING SYSTEMS

Modern railroad lines use block signaling, first used by the New York & Erie Company in 1849. No train is allowed to enter a block, or section, of track until the previous train has left. The early system used colored flags and signs operated by signalmen by the track, but an automatic system was invented by Thomas Hall in 1867.

One rail in each block carries a small electric current. When a train arrives in the block, the current runs through the wheels to the other rail, causing the signal to change. When the train leaves the block, the current is confined to one rail again, making the signal move to the next color setting.

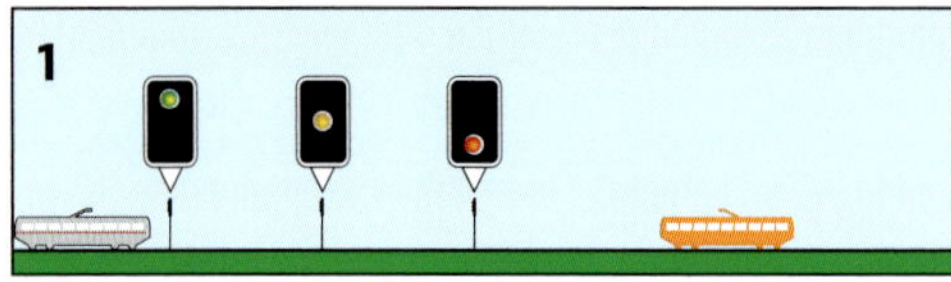

1 The gray train has a green signal so the driver can travel into at least the next two blocks.

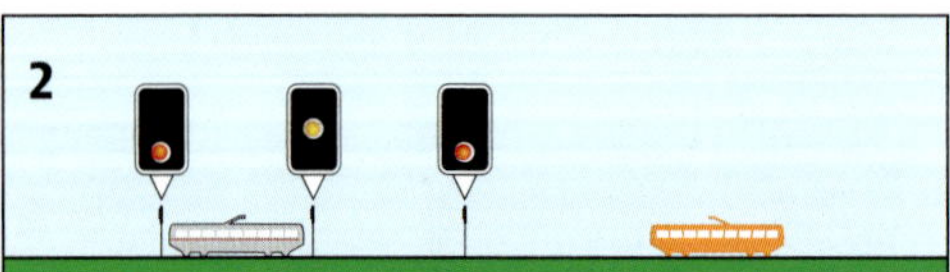

2 As the train passes the first signal, the green light turns red. The next light is yellow, warning the driver that the next signal is red.

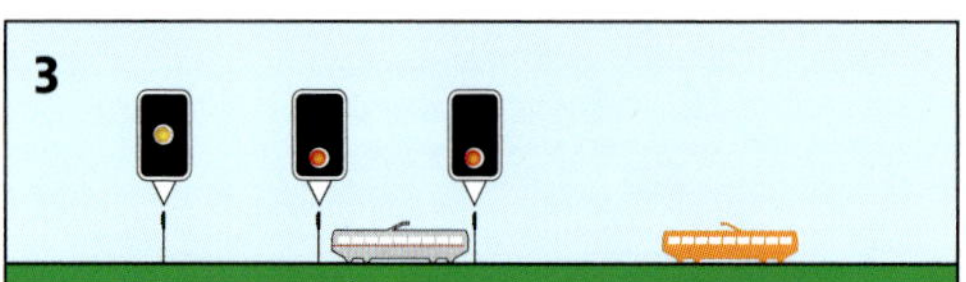

3 As the train passes the second light, it turns red. The first light turns yellow. Since another train is in the block ahead, the third light is red, signaling the driver to stop.

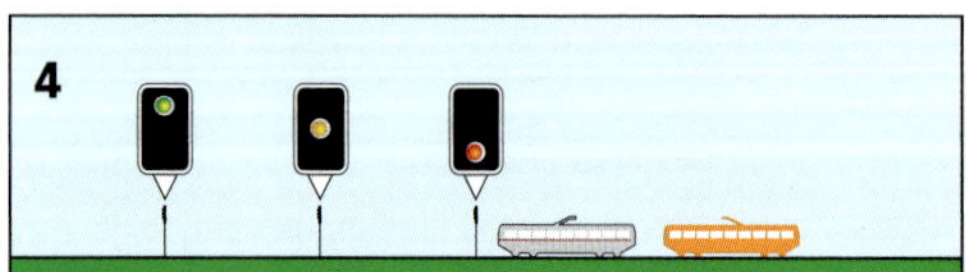

4 The driver passes the red by mistake, and an alarm sounds in the cab as a warning. The first light is green again since the two blocks ahead are now clear.

SCIENCE WORDS

Electromagnet: A magnet that can be turned on and off.
Levitation: To float in the air when an upward force is balancing out gravity.
Transmission: The mechanism that carries the energy produced by the engine to the wheels, making them turn.

CLIMBING HILLS

Three systems have been used for taking passenger cars up steep gradients. One is the cable car in which each car is fitted with a clamp that grips onto a moving cable. The famous cable cars of San Francisco first ran in 1867 and were designed by Andrew Hallidie (1836–1900).

The second is the funicular railroad, which has two cars fixed to opposite ends of a cable. As one car goes up the slope, the other comes down. An engine at the top controls the cable, but little power is needed, because the two cars almost balance one another. The first funicular was built in 1879 to carry tourists up Mount Vesuvius in Italy.

Finally, the most widespread system is the rack railroad, which has interlocking teeth on the car wheels and rails. Developed by Swiss inventor Niklaus Riggenbach (1817–1899) in 1862, the original design could cope with gradients of about six percent, but later versions were built on much steeper slopes.

could not cope well with the strain of getting a heavy train underway. In the same year, however, the first diesel-electric unit was built in Sweden. It used a diesel motor to generate electricity, which then powered the locomotive. This was the system that would eventually replace steam, since it combined high speeds with low fuel consumption and low maintenance. The world's first regular scheduled diesel-electric freight service was opened on the Santa Fe line in the United States in 1940.

A metro train runs along an elevated track through the Hague, Netherlands. It was built to connect older tram and train tracks in different parts of the city.

In Budapest, a funicular railroad carries tourists up the hill of Buda to the castle.

Diesel engines with mechanical transmission have been used successfully in lightweight rail cars, which combine power and passenger accommodation in one unit. Some diesel-hydraulic systems have also been developed, especially in Germany. In these locomotives, the diesel engine is used to generate hydraulic (liquid) pressure, and this rotates a turbine to drive the wheels.

Freight Systems

Recent inventions have centered on freight handling, safety, and high-speed trains. Highways built in the last half of the 20th century took many customers away from the railroads. Rail services fought back by introducing piggyback and container services. Carrying containers by rail meant that freight could pass from ship to train to truck without ever leaving its own container. Piggyback services carry road trucks with their loads of freight. The trucks drive away from the rail terminus to their final destinations.

ALTERNATIVES TO STEAM

An experimental battery-powered locomotive ran in the United States in 1839. But the first really successful use of electric power was in a locomotive shown at the Berlin Fair in the summer of 1879. Developed by German engineers Werner von Siemens (1816–1892) and Johann Georg Halske (1814–1890), it proved that electric-powered trains were a practical possibility. A street railroad opened in Germany in 1881, and a small electric railroad in Ireland in 1884. The London subway system began using electric locomotives in 1890.

Many commuter routes rely on diesel-electric trains. Electricity generated by the engine not only drives the train but also powers doors, lights, and brakes.

CENTRALIZED CONTROL

Block signaling is used when trains are following one another along the same track, but modern rail networks also use Centralized Traffic Control. A display in the control center shows the position of each train, and the color of every signal. Rolling-block signaling is another new innovation. There are no signals beside the track; instead, they shown in the driver's cab. This is useful on busy networks, such as subway systems.

FACTS AND FIGURES

This table lists the countries with the longest railroad systems, and the date the first track was laid.

Country	Date	Length	Gauge
United States	1830	136,729 miles (220,044 km)	4 ft 8½ inches (1,435 mm)
China	1880	98,798 miles (159,000 km)	4 ft 8½ inches (1,435 mm)
Russia	1837	65,244 miles (105,000 km)	5 ft 0 inches (1,524 mm)
India	1853	42,616 miles (68,584 km	5 ft 6 inches (1,676 mm)/3 ft 32⁄5 in (1,000 mm)
Canada	1836	30,709 miles (49,422 km)	4 ft 8½ inches (1,435 mm)
Argentina	1857	22,970 miles (36,966 km)	5 ft 6 inches (1,676 mm)/3 ft 32⁄5 in (1,000 mm)
Germany	1854	20,723 miles (33,351 km)	4 ft 8½ inches (1,435 mm)/3 ft 6 in (1,067 mm)
Australia	1845	20,610 miles (33,168 km)	4 ft 8½ inches (1,435 mm)
Brazil	1835	18,527 miles (29,817 km)	4 ft 8½ inches (1,435 mm)
France	1832	17,077 miles (27,483 km)	4 ft 8½ inches (1,435 mm)

HIGH-SPEED TRAINS

The French electric TGVs currently hold the record for the world's fastest wheeled train. In 2007, a TGV achieved 357.2 mph (574.8 km/h). The Chinese Harmony locomotive operates routinely at the fastes speed. It runs at 217 mph (350 km/h). Both trains are powered by electricity, collected from overhead cables using a pantograph. This metallic arm extends upward to touch the cable and acts as a conductor, allowing electricity to pass from the cable to the train. High-speed locomotives have powerful electric motors that drive the wheels. Air vents keep the motors cool. The wheels are mounted on swiveling trucks, and suspension springs ensure a smooth ride. The typical train has eight passenger cars coupled to a motor unit at each end.

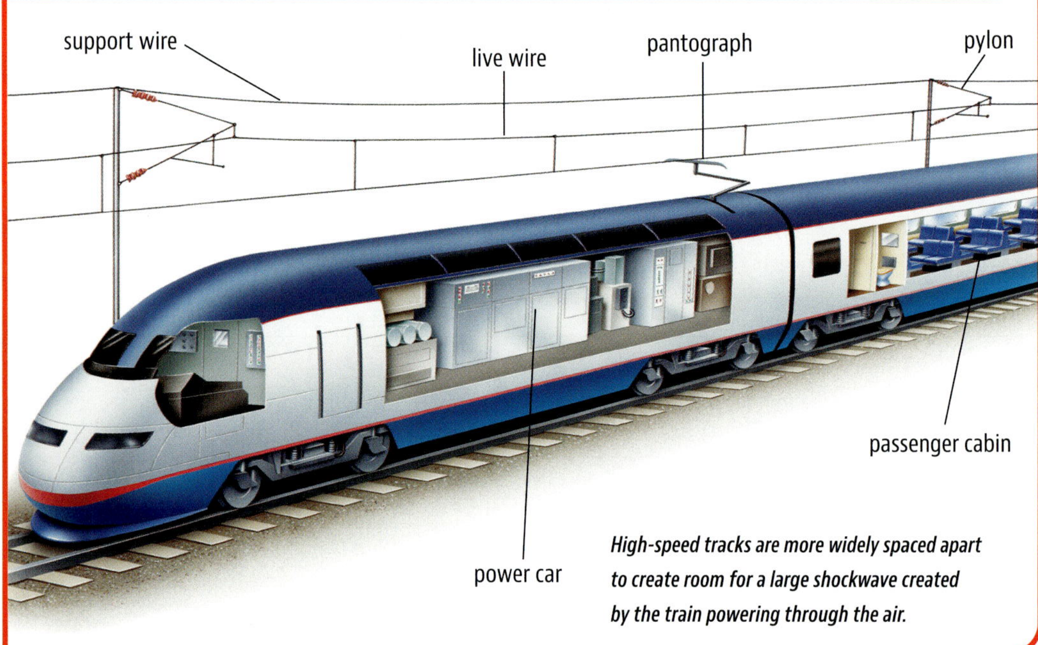

High-speed tracks are more widely spaced apart to create room for a large shockwave created by the train powering through the air.

Freight handling is speeded by computer systems that plan routes and monitor progress, aiming to minimize costs and maximize speed of delivery.

"Bullet" Trains

Competition with inexpensive airline flights led to the development of high-speed trains. In 1964, the electrically powered "bullet trains," were introduced in Japan. Japanese high-speed trains operate at speeds of up to 200 mph (320 km/h). In France, gas-turbine engines were developed. These used hot gases turning turbines to generate electric power (or sometimes drive the wheels directly), but due to high oil prices they were uneconomic. The French success has been the electric TGV (Train à Grande Vitesse)—a permanently coupled, lightweight train, running on a specially built track.

Maglevs

A train running on a single central rail is called a monorail. Monorail maglev trains are currently the fastest type of locomotive. Maglev stands for magnetic levitation.

MAGLEV SYSTEM

1 Suspension Electromagnets, fixed to the outside of the train, are positioned so they lie underneath the track's suspension rails. When the power is on, the electromagnets are attracted upward toward the rails but are not quite powerful enough to reach them, because of the train's weight. This lifts the train, so it floats around the track.

2 Propulsion A linear motor is fitted under the train. It consists of a line of magnets (linear motor coils) that can switch between poles (shown as red and blue). The magnets are switched back and forth, creating a surge of magnetic field flowing along the train's length. This produces another magnetic field in the reaction rail that runs along the center of the track. The two fluctuating magnetic fields push and pull against each other, forcing the train along.

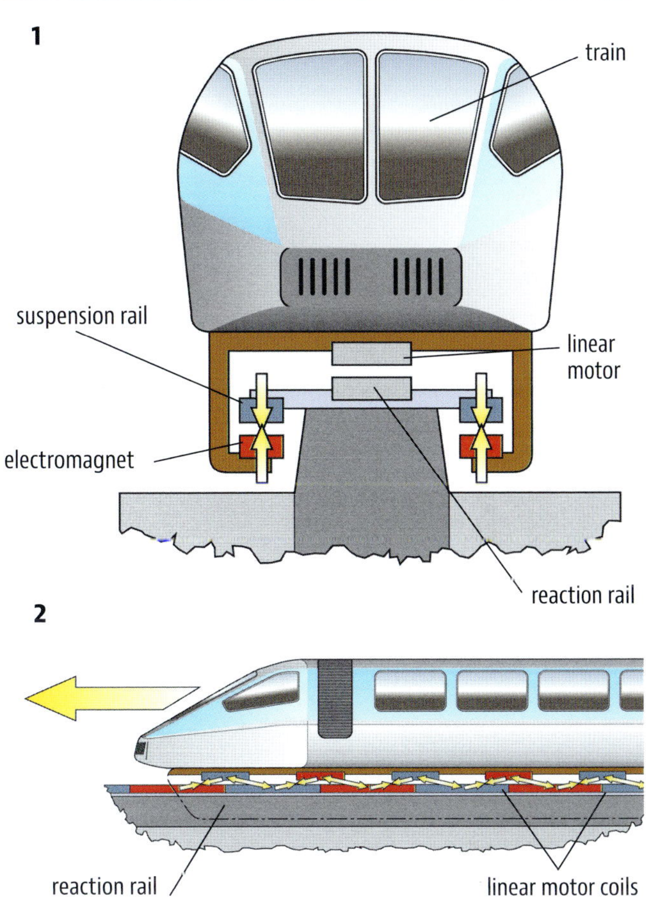

The trains have no wheels but float above the track—and so achieve great speed. The fastest train on record is a Japanese maglev, which managed 375 mph (603 km/h) in 2015.

The country with the largest network of high-speed trains is China, accounting for two thirds of all the high-speed railroads in the world. The Beijing-Guangzhou high-speed line alone is 1,428 miles (2,298 km) long

A high-speed maglev train connects Shanghai, China, with the city's main airport. It can reach a speed of 268 mph (431 km/h).

ON THE ROAD

Gasoline-driven vehicles took to the road more than 100 years ago, but the first road vehicles had appeared long before that. These early vehicles were driven by steam.

The world's first self-propelled road vehicle was a steam carriage built in 1769 by Nicolas-Joseph Cugnot (1725–1804). His immense vehicle had mixed success, and steam engineers focused on building railroad locomotives or powerful stationary engines. However, Cugnot's carriage inspired others to develop steam engines small enough to fit into a road vehicle.

In the United States, Nathan Read (1759–1849) and Apollo Kinsley both ran steam vehicles in the 1790s. By 1800, there were steam buses in Paris—but only for a short time. In England, a Cornish mining engineer, Richard Trevithick (1771–1833), built a machine with driving wheels 10 ft (3 m) in diameter. But it was broken up when Trevithick decided that railroads were a better prospect for transportation in the future. Later, however, steam machines began to reappear on city streets.

STEAM CARRIAGE

Early steam engines were far too big to use for transportation. But eventually, in 1769, French military engineer Nicolas-Joseph Cugnot succeeded in building a three-wheeled carriage, which was demonstrated in Paris. Traveling at less than walking speed, it crashed into a wall and then literally ran out of steam after just 15 minutes. The military authorities ordered an improved version to pull cannons—but then never used it.

The boiler on Cugnot's carriage was a large metal vessel at the front. It was this part of the vehicle that crashed into a wall during the world's first automobile accident.

The first practical steam truck worked in Glasgow, Scotland, in the 1870s. Steam delivery trucks started operating in Paris in 1892. Steam trucks were used by some breweries and coal merchants right through to the 1920s. But it was in the countryside—in agriculture and heavy traction—that steam power really made its mark in the form of early tractors. In the later 19th century they became a familiar sight, pulling and powering early farm machinery, especially on the flat prairies of North America. In many countries steam tractors became a feature of the large traveling circuses of the time. The vehicles hauled the showmen's wagons by day and then drove carousels at night.

Building a Car

Two German inventors were critical in developing the modern automobile and neither, initially, was interested in four wheels. Gottlieb Daimler (1834–1900) put his prototype engine into a motorcycle, while the other German pioneer used a tricycle.

In doing so he made what is now generally accepted as being the first real automobile. That other inventor's name was Karl Benz (1844–1929).

Like Daimler, Benz began his career as a maker of stationary gas engines. His advanced designs proved successful, and his business was thriving, so it was a shock to his associates when he announced that he wanted to put money into building a motor vehicle. But Benz was determined, and soon converted one of his engines to run on benzene (a chemical compound of carbon and hydrogen). All he needed was a vehicle to put it in. He decided that three wheels would give stability and simplicity, and he drew on cycle technology to keep down the weight of the vehicle.

The result, in 1886, was a spindly, uncomfortable road machine, but it worked very well and was reliable. It made journeys of several miles possible. Karl Benz worked to improve the design and began to sell his automobiles to the public. This was the birth of the motor industry.

PUBLIC TRANSPORTATION

Walter Hancock (1799–1852) tried to use the new steam carriage to challenge the supremacy of horse-drawn omnibuses on the streets of London, England. Beginning in 1831, he built nine impressive machines, culminating in a 22 seater that could travel at 20 mph (32 km/h). His vehicles were called "Enterprise." During the building of one of these vehicles, an engineer died from fright when a boiler exploded. But road toll charges for steam-powered vehicles increased, and by 1836 Hancock had been driven out of business.

One of Walter Hancock's "Enterprise" omnibuses driving through a London park, to the alarm of onlookers.

DAIMLER MOTORCYCLE

In the 1860s and 1870s, as steam engines became smaller, lightweight steam cars were designed in a number of countries. One experiment in 1868 even used a simple velocipede bicycle of the time in an attempt to make a steam motorcycle. But the true ancestor of the modern motorcycle was built in 1885 by Gottlieb Daimler. Throughout the 19th century, inventors had been working toward an effective internal-combustion engine. Daimler built a benzene-powered engine and needed a vehicle to test it on. He constructed a crude, wooden-framed motorcycle that anticipated later designs in having the engine mounted in the middle and driving the rear wheel. But he also added small outrigger wheels that could be lowered to keep the machine on an even keel. With no suspension and tireless steel-rimmed wheels, riding it must have been a difficult and exciting experience. Daimler soon abandoned the motorcycle and turned his attention to the automobile—with much greater success.

Making Progress

Automobile production quickly spread to other countries, and design developed rapidly. The year 1895 was a particular turning point. The first automobile with a fully enclosed engine and a body appeared: this was the Panhard-Levassor, built in France. The first brake shoe was developed, and the first direct driveshaft transmission system was designed in the same year.

Commercial vehicles also appeared in 1895. The first gasoline-driven motor bus went into service, over a 9 mile (15 km) route, in North Rhineland, Germany; the first gasoline-engined van was put into production; and the first prototype gasoline-engined truck, also a Panhard, was tried out in Paris. Other improvements soon followed: glass windshields in 1903, fenders in 1905, and rearview mirrors in 1906. Private cars remained hobbies for the wealthy, but in 1913 a mass-produced car by Henry Ford was to change that.

Early steam tractors, complete with smokestack, resembled small rail locomotives, although with four wheels.

THE EARLY AUTOMOBILE

The first automobiles owed a great deal to the horse-drawn carriage. In fact, many were simply coaches with an engine—one reason why they were known as "horseless carriages." Even specially built cars were usually made by traditional coachbuilders and so had the same large wheels, distinctive body, and high driver's seat. Light wheels and slow speed meant that the first cars could be steered with a small tiller on an upright column in the middle of the car. The engine settings had to be adjusted constantly using levers on the steering column or a control nearby. Speed was controlled by moving a lever back and forth. There was no brake pedal in the early car, just a hand brake that clamped brake pads against the wheel rims (like modern bicycle brakes) or pulled a belt tight around the axle.

Weaker drivers could not pull on the brake lever hard enough to stop a speeding car. These first motor vehicles often jolted along on spoked bicycle wheels or on heavy wheels adapted from carts with wooden spokes and solid rubber rims. By 1895, however, many were beginning to use air-filled pneumatic tires, which gave a more comfortable ride, although punctures were frequent. The first pressed steel wheel was introduced in 1910. In many early cars the power of the engine was transmitted to the wheels by a chain or even a leather belt. But in 1895, the Renault company championed the idea of a rigid metal drive shaft linked to universal joints through which power can be passed in any direction. This system is still used today.

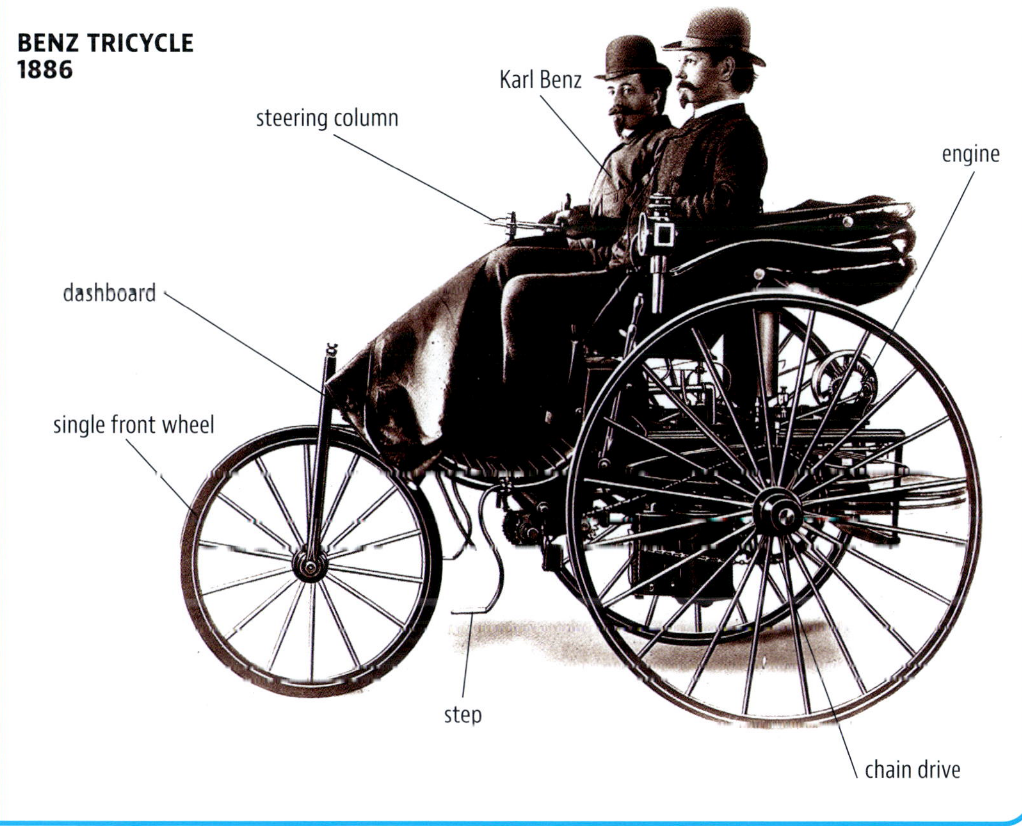

RACING MOTORBIKE

Racing motorbikes have the most advanced motorbike design. Electronic ignition starts the engine; older models were kick started. The engine is a smaller version of a car engine, with cylinders that burn fuel inside and push pistons down. These rotate a crankshaft. A chain drive transmits the power of the engine via a gearbox to the back wheel. The engine is cooled by liquid that circulates around the cylinders and through a radiator, where air is drawn in to cool the liquid. Burned gases pass out through the exhaust.

The rider uses hand and foot levers to brake, a twist-grip on the right handlebar (throttle) controls the engine's speed, and one on the left handlebar controls the clutch (which engages or disengages the engine from the transmission system). The brakes are disk brakes. The rider wears a suit made from tough leather with extra protection on the back. In case of an accident, the suit allows the rider to slide over the ground and slow down gradually—hopefully avoiding serious injuries. A helmet protects the head.

More Inventions

Thousands of small inventions continued to make automobiles more powerful, more comfortable, and easier to control. Brakes, for example, have changed dramatically from the earliest examples. These cars had no brake pedal, only a parking brake lever. The brakes themselves were usually just pads that pressed on the rim of the wheel. Whether the brakes actually stopped the vehicle depended on the strength of the driver.

By the early 1900s foot-operated drum brakes were the standard system. The brake pads were semicircular brake shoes inside a drum fixed to the inside of each wheel. The drums kept off rain and mud. But the brakes still needed regular adjusting to maintain equal pressure on each wheel.

Many of the first road cars were built in France by companies, such as Peugeot and Renault. These manufacturers are still producing automobiles today.

MICHELIN TIRES

The first automobile to be fitted with pneumatic tires was a Peugeot with a four horsepower Daimler engine. It competed in the Paris-Bordeaux race of 1895. The tires had to be changed 22 times during the 750 mile (1,200 km) race, even though each tire was fitted to its wheel with 20 nuts and bolts. The driver, Edouard Michelin, went on to found the successful Michelin tire company.

Cars should carry a jack, a simple lifting device that acts as a lever and is used to raise the car and tire off the ground so it can be changed when it is damaged.

WHO INVENTED THE AUTOMOBILE?

People have argued long and hard over who invented the automobile. Some say it was the Belgian Étienne Lenoir, who drove a gas-engined carriage as early as 1862 and even sold one to the Czar of Russia. Some people argue for the Austrian Siegfried Marcus, who fitted an engine to a handcart in about 1870. Another contender is Frenchman Edouard Delamere-Deboutteville, who in 1883 tried putting a gasoline engine into two different chassis (both proved too fragile). But it was the German engineer Karl Benz who was the first man to build a vehicle with an internal-combustion engine and turn it into a practical commercial proposition.

On modern production lines, cars bodies are assembled largely by robots. Later, human workers fit the engine and the many extra smaller components.

FACTS AND FIGURES

• When automobiles first appeared in Britain they were subject to the same laws that were applied to steam traction engines. Officially, someone had to walk in front carrying a red flag to keep its speed at 4 mph (7.2km/h).

• The dashboard, or instrument panel, got its name because on horse-drawn carriages there was a board in front of the coachman to stop him being "dashed" by stones thrown up by the horses' hooves.

• The first automobile to be used in an election campaign was the Mueller-Benz lent to American Democratic presidential candidate William Jennings Bryan in October 1896.

• By 1913, the United States had become the first country to have more than a million cars.

• Current estimates are that there are 1.4 billion vehicles in the world, including 363 million trucks and buses.

Braking Systems

By the 1930s, automobiles were becoming much faster and heavier and drivers needed more assistance to control them, particularly to stop them. Engineers came up with a solution by inventing the hydraulic brake system. The brake pedal and brakes were linked by tubes filled with oil, instead of metal cables.

However, by the 1950s automobile speeds and weights had again overtaken the braking system. The power-braking system was devised, which again proved to be a solution. Brakes were still hydraulic, but a vacuum system provided the power, not just the force applied to the brake pedal.

Finally, in the 1970s disk brakes began replacing drum brakes—usually just on the front wheels, with drums still on the rear. Disk brakes had been invented back in 1902, but had been too expensive to make. Now, they were dependable and essential.

Future Technology

ABS brings us to the present and future of automobile innovation—for it was the first major invention made possible by putting

HENRY FORD AND THE MODEL T

Businessman Henry Ford (1863–1947) set up his Ford Motor Company in June 1903. At first he built his cars in the usual costly way, as individual hand-built items. The total world output of cars in that year was under 62,000 vehicles—almost half of them built in France. Then in 1908, Ford introduced the revolutionary approach that was to make his name: a standardized car, the famous Model T. This had simplified spare parts and, to keep costs down, a production line on which each worker performed a single task as the vehicles being assembled moved by on a conveyor belt. Between 1908 and 1927, 15 million Model T Fords were built—the equivalent of over 14,000 a week. By 1920, half the cars in the world were Fords.

Henry Ford famously said, "Any customer can have a car painted any color that he wants, so long as it is black."

SOCIETY AND INVENTIONS

A Car-based Society

The motor vehicle has been one of the greatest forces for change in the 20th century. The mobility it allows has affected every aspect of people's lives: where they live, where they work, where they shop, where they go for leisure. The result has generally been far greater choice and freedom. But a more mobile society has also, perhaps, meant less community spirit. It has made living more compartmentalized: there are separate areas for living, areas for working, areas for shopping, and so on. At the same time, the needs of the automobile have radically changed our surroundings and environment: more and more roads leading to increased traffic, more parking spaces, and more pollution caused by gasoline and diesel fumes.

The cloverleaf interchange, in which drivers move from one highway to another without stopping, was invented by Arthur Hale in 1916. Interchanges such as this take up a lot of space, but allow cars to keep moving.

THE ABS SYSTEM

The natural reaction in an emergency is to put the foot firmly on the brake pedal and keep it there. This can make the wheels lock and the automobile skid. It will stop, but a vehicle slows best when its tires are gripping the road, so friction between tire and road is high. As soon as the wheel skids, that grip is lost. In the 1970s the Bosch company of Germany developed ABS (Antilock Braking System) to solve this problem. ABS uses sensors on each wheel that send information to a computer. When a locked wheel is detected, the computer can release and reactivate the brakes 15 times a second. The driver can still steer the braking vehicle, making it possible to drive around obstructions.

computers into cars. Most cars have traction control, another computerized technology that prevents wheelspin when accelerating. But perhaps more of the future may be glimpsed by looking at racing cars.

Racing Technology

In the 1990s, international Formula One racing cars had reached a high level of engineering sophistication. They had ABS and traction control, and computerized technology was introduced to help perform other functions. Fully automatic transmission ensured the right gear at the flick of a switch.

In Formula One cars, computerized traction control automatically prevents unwanted wheel-spin when accelerating, and four-wheel steering (rear wheels turn, as well as the front

SOCIETY AND INVENTIONS

The "People's Car"

The Volkswagen Beetle, one of the world's most recognizable and popular automobiles, illustrates the importance that politics and society can have on technology. In 1938, the Nazi leader Adolf Hitler launched the Volkswagen VW—the "People's Car"—a low-cost mass-produced car designed for ordinary drivers. The VW company was set up by the government. At the same time, the Nazis were developing Germany's autobahn system, a network of roads that became a model for modern expressways. After World War II, a U.S. advertising campaign nicknamed it the Beetle, for its rounded shape. Production of the Beetle car in its original design ended in 2003.

The original design of the Beetle is no longer manufactured, but these classic cars are still maintained—and raced—by fans.

A disc brake system before the wheel is mounted.

The brakes on many modern high-performance cars are left showing.

BRAKING PRESSURE

A car's brake pedal is linked to the brakes by pipes filled with oil. Pressing down on the pedal moves a piston that puts pressure on the oil, and this pressure passes to the brakes. This system instantly puts equal pressure on left and right brakes, thanks to the principles of hydraulics—the laws about the behavior of fluids. When a liquid is in a closed container, the pressure throughout the liquid is always equal. If you press a liquid at one end, the force immediately passes to the other. Since a car's brakes are interconnected, they form one "container," so the pressure on each side is always the same–an important safety consideration.

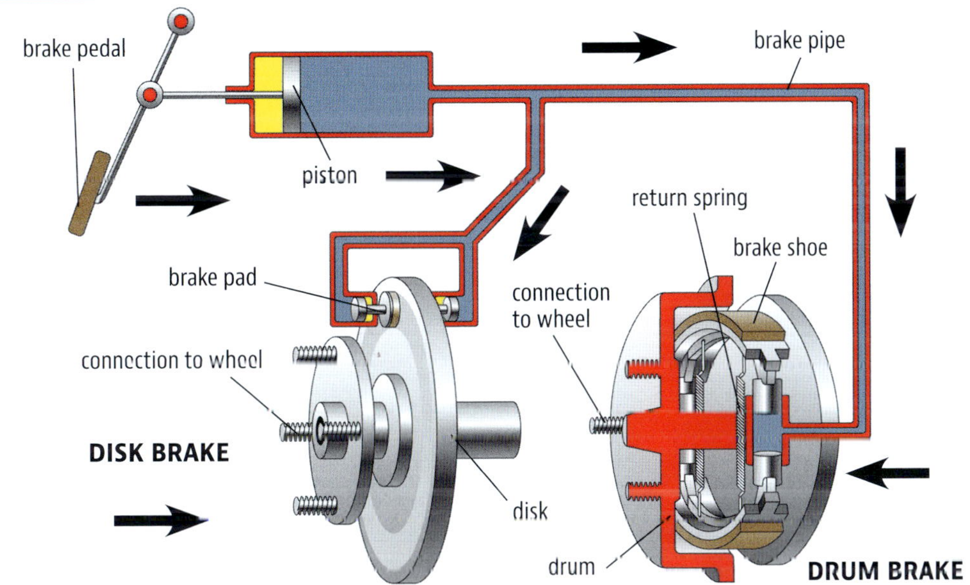

Modern cars have disk brakes on the front wheels, and drum brakes on the rear.

In a drum brake, "shoes" press against the inside of the drum. Springs pull them away when the brake is released. A disk brake has pads that pinch on either side of a disk to slow the spinning wheel.

Formula One cars move at around 200 mph (320 km/h) during a race. At this speed they would lift off the track and be hard to control if it were not for an upside-down airfoil "wing" at the rear, which pushes them back down.

SELF-DRIVE AUTOMOBILES

The next generation of automobiles may be able to dispense with a human driver altogether. Developments in global positioning systems (GPS), sensors, and computers mean that self-drive cars can effectively navigate and maneuver in almost any environment. Self-drive cars, also called autonomous cars, use cameras and sensors to build a picture of their surroundings. This information is fed to an on-board computer, which tells the car to brake, accelerate, or change direction. The problems involved with self-drive cars are not so much the basic technology as ethical issues. For example, who would be responsible in the event of an accident.

Self-drive cars use data from sensors combined with artificial intelligence to navigate and avoid obstacles and other road users.

ELECTRIC CARS

As the cost of drilling for oil rises and the links between burning fossil fuels and climate change become more obvious, automobile manufacturers are looking to alternative fuels. Hydrogen cars use fuel cells, which work a bit like batteries. When hydrogen from a tank combines with oxygen from the air, it releases energy and creates nothing more polluting than steam. Although hydrogen cars have been tested, fuel cell cars are expensive and currently there are no systems to make and distribute lots of hydrogen.

The main technology that is now considered the future of the automobile is the electric car, powered by rechargeable batteries. These cars are efficient: they have good acceleration and are also quiet. Many countries have decided to phase out cars powered by gasoline or diesel fuels in the next two decades, and sales of electric cars are steadily increasing. As the technology of electric cars has improved, the number of miles that can be driven before it is necessary to recharge the battery has come down. The range of most electric cars before they need to be recharged is around 300 miles (500 km).

Some people are concerned by the time it takes to recharge the battery. While it takes 10 minutes to refill a car with gasoline at a service station, it may take four or more hours to recharge a battery at a charging point. Something else for drivers to consider is that the number of charging points available in most cities, towns, and villages is limited.

Many of the electric-powered cars on the road today are hybrids. They use both gasoline and electric motors. Electricity powers the car at low speeds, before the gasoline engine kicks in. This does not, however, eliminate the need for fossil fuels.

FORMULA ONE

Formula One cars draw on the latest innovations in aerodynamics, material science, and engine design. Race cars have used front and rear wings for over 50 years. These work the opposite way to a plane's wings, making a downward-pushing force that keeps the car pressed to the track as it corners at high speed. Today's wings use a much more complex design to produce maximum down force with minimum air resistance, and they're built from strong, lightweight aerospace materials such as carbon fiber.

Concerns about the environmental impact of Formula One cars have led to improvements in engine power. Today's cars have hybrid engines and race on E10, an eco-friendly blend of gasoline and ethanol. Another important development is the use of energy-saving braking. Normally when a car brakes, the kinetic energy it has (due to its movement) is wasted when it turns to heat in the brake pads. Energy-saving brakes store this energy for reuse, either in a heavy drum, rotating at up to 64,000 rpm, or in a battery. When the car brakes, up to two thirds of its energy is stored as the car slows down. This saves fuel, and the driver can reuse the stored energy next time they accelerate.

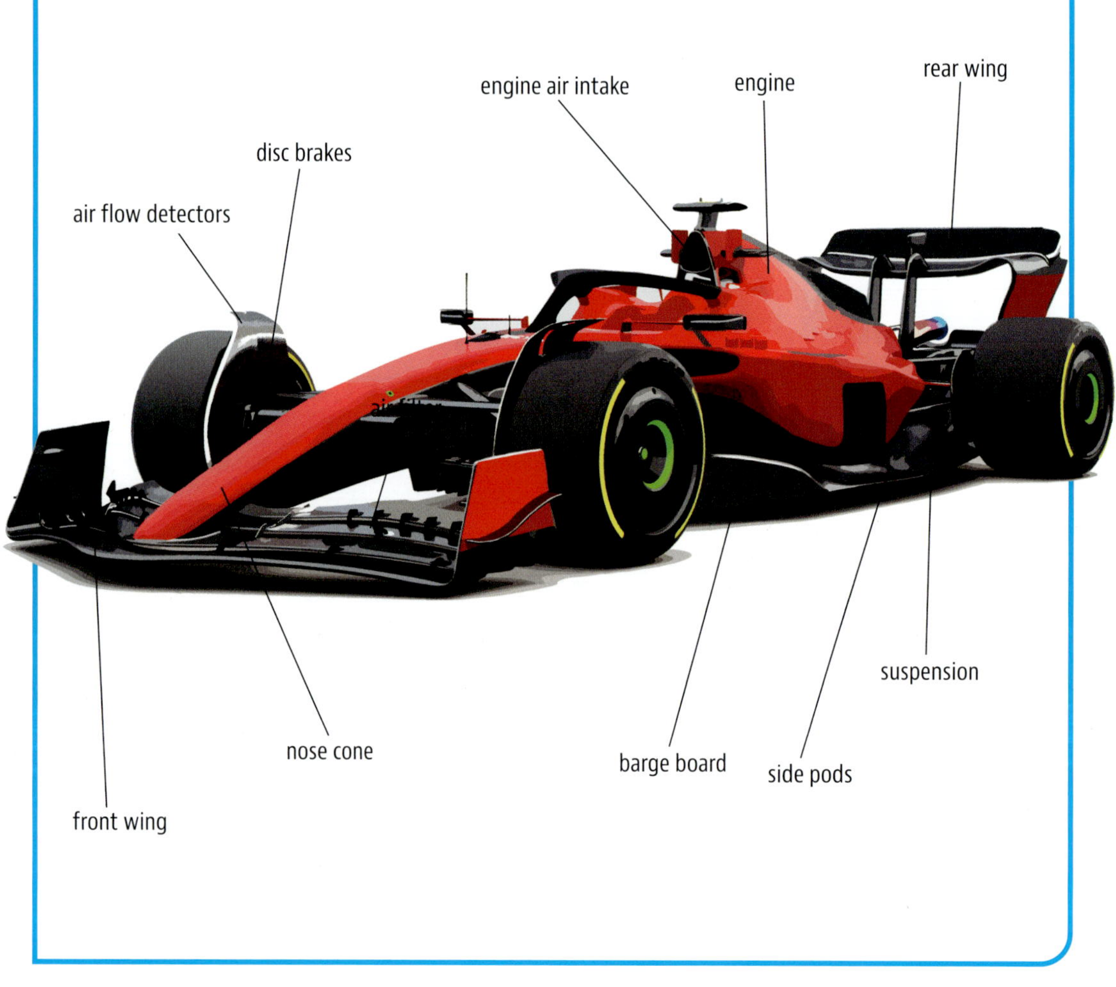

Electric scooters are used in many cities worldwide. They are regarded as a clean, convenient, and cheap form of short-range transport within urban areas.

wheels) helps improve stability. In addition, on-board computers store information from each lap of the circuit. Timing and cornering speeds are monitored so that every gear shift, acceleration point, braking force, and suspension movement are monitored to help the driver achieve the fastest lap.

Every season, the technical specifications and regulations for Formula One cars change, and teams of engineers work on developing cleaner engines while retaining all the power needed to win races. The technology used in hybrid and fully electric racing cars has fed back into the automobile and aerospace industries, and is now being used in road cars and airplanes as a matter of course. Formula One is working to be carbon-neutral by 2030.

PUTTING DRIVERS IN THE PICTURE

Today, most automobiles have an array of high-tech electronics and computerized control systems. In recent years, for example, driver information display has been revolutionized by digital technology, such as satellite navigation maps. Analog (clock dial) displays have been retained for the speedometer, fuel levels, and rev (engine revolution) counter because drivers find it easier to take in changing analog information at a glance. Other functions, such as the stereo, heating, and air conditioning can be controlled from a single touchscreen.

Navigation systems pick up signals from satellites to calculate the car's position. A computer then calculates the best route to the destination.

SHIPS AND BOATS

People have traveled by boat for many thousands of years, and ships have played a crucial role in the spread of civilizations throughout history. Today's vessels are the largest forms of transportation ever made.

There is no real way of knowing when boats were first used, but it is known that the ancestors of the Aboriginal people used boats to reach Australia at least 50,000 years ago. It is also known that other people spread through the Pacific islands around the same time, meaning that they must have had ocean-going craft. The first people to arrive in the Americas about 14,000 years ago probably traveled by sea down the west coast.

Ancient people would have seen logs floating down rivers and used them to make rafts—and then dugout canoes. Canoes were also made by stretching animal skins over a

DUGOUTS

The oldest remains of boats that have been found are around 10,000 years old. These were hollowed out logs that could be propelled and controlled by poles or paddles. They enabled people to cross calm stretches of water and to explore new lands.

A simple canoe cut from a tree trunk.

Modern luxury speedboats use a combination of streamlining in their hulls, responsive propellers, and powerful engines to reach high speeds. At top speed much of the hull is raised above the water to minimize drag.

light frame of bent wood and binding them together with strips of leather. A skin canoe dating from around 4500 BCE was found on the Baltic island of Fünen. Skin canoes are lighter than dugouts, can be carried from river to river by one person, are easier to handle, and cope in surprisingly rough water.

One fascinating boat of this kind is the *quffa*, used on the Tigris and Euphrates rivers in Iraq and dating back thousands of years. Quffa are round canoes up to 18 ft (5.5 m) across that can carry up to 20 passengers. They are made by sealing basketwork with tar.

Right across the world, from Australia to North America, people learned how to make canoes from tree bark sewn together with tree roots and sealed with resin. These bark canoes proved both light and strong.

A replica of the Greek trireme Olympias*.. The ship was powered by 170 oars and had two sails. It had a top speed of about 9 knots (17 km/h). Speeding triremes were used to ram enemy ships in battle.*

PLANK BOATS ON THE OPEN SEA

By lashing planks together, boats can be built much bigger than either a dugout from a single tree or a skin canoe. In 2000 BCE, the shipbuilders of ancient Egypt were making ships over 100 ft (30 m) long by interlocking 1,000 or more small planks of wood and lashing them together with tough grass rope. By 1000 BCE, the traders of Phoenicia (now Lebanon) were venturing out of the Mediterranean and braving the vast open spaces of the Atlantic in stout sea-going ships made from long planks of cedar held together with wooden ribs.

Planking—which could be either edge-joined or overlapped—not only made it possible to make big, light, seaworthy ships, it also allowed boats to be shaped so that they cut smoothly and swiftly through the water. Nowhere was this demonstrated more clearly than in the long, low, elegantly shaped longships made over 1,000 years ago by the Vikings, which carried them from Scandinavia as far as North America.

Traditional dhows have triangular lateen sails. These sails are more flexible than square rigged sheets. Using this type of sail makes it easier to "tack" into the wind.

SAIL POWER

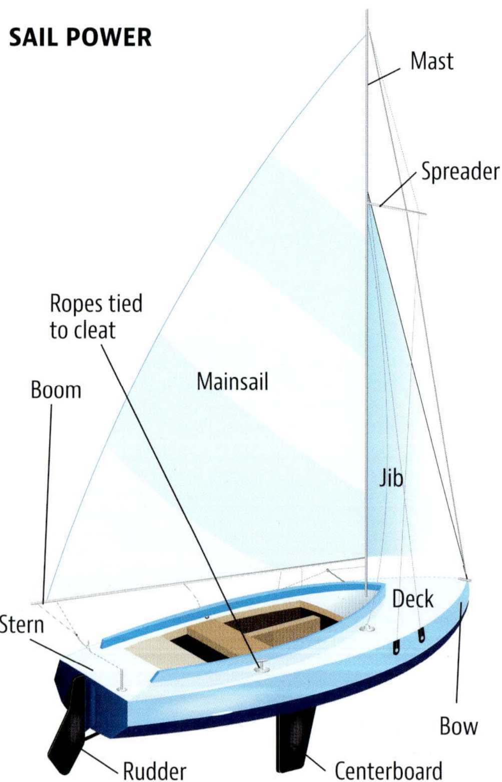

A sloop is a small sailboat. It has two triangular sails and can be controlled by just one person. Sloops are only suitable for sailing in coastal waters close to land.

Native American Algonquins and Iroquois both made canoes from tree bark. The Algonquins used white paper birch bark, which proved to be the best material, but the Iroquois used elm bark. These bark canoes were the inspiration for today's sport canoes made from fiberglass and plastic. All these early craft were powered by hand, either using a pole that was pushed along the bottom of the river, or with paddles dipped into the water. Sails that used the power of the wind to propel boats were probably first used about 6,000 years ago.

HOW BOATS SAIL AGAINST THE WIND

Sailing ships can sail against the wind because when sailing in that direction the wind does not push the sail so much as suck it. The sail acts like an airfoil (the shape of an aircraft wing). As the wind blows over the curve of the sail, it speeds up and the pressure drops on that side, creating a lift force—just as above a wing. This force pulls the boat forward. By zigzagging at the right angle, first one way then the other—a technique called tacking—a boat can make headway almost directly into the wind. Some modern yachts can sail as close as 39 degrees to the angle of the wind (0 degrees is straight into the wind).

CLIPPERS AND TALL SHIPS

The pinnacle of sailing ship technology was reached in the clippers of the 1800s. These ships got their name because they tried to "clip" time off the voyage as they raced to be first to get their cargoes of tea from India and China to the markets in Europe and North America.

Designers reduced the hull to a sleek minimum and mounted huge areas of sail on tall masts to make the most of the wind. Clippers like these could reach speeds of well over 30 knots (35 mph, or 56 km/h). In 1866 the clippers *Taeping*, *Serica*, and *Ariel* raced 15,970 miles (25,700 km) from Fuzhou in southern China to London in just 99 days.

Like other tall ships, clippers had a complex arrangement of square sails and triangular lateen sails. The sails were made from canvas (heavy cotton). Square sails were hung across the width of the ship from crossbars, known as yardarms. The square sails could pivot on horizontal booms to take maximum advantage of different wind directions.

The lateen sails were hung along the length of the ship and were there to pick up cross winds on either their front or back surface. This combination allowed clippers to sail almost into the wind, so they were less at the mercy of the wind direction. The huge area of sail also allowed them to take advantage of even the gentlest breezes.

Tall ships—so called because of their tall masts, which are almost as high as the ship is long—still sail the oceans today. Many are used as training vessels for young naval cadets to "learn the ropes"—the basics of sailing.

The first known picture of a sail dates from 3500 BCE and comes from Egypt. The sail is simply a square of cloth hung from a spar mounted on a mast. These earliest sailing boats had square sails and depended on a following wind. However, by 200 CE sailors in the Mediterranean were using triangular lateen sails, much like those used on Arab dhows, which were better suited to sailing into the wind.

Oceangoing Vessels

For 1,000 years or more merchant ships changed little. They were steered by an oar over the side and rarely had more than two sails. But around 1200 the steering oar was replaced on European ships by a hinged rudder over the stern (the rear of the boat). This Chinese invention made the boat much more maneuverable, allowing progress even when the wind was against it. A mixture of square and lateen sails meant boats could sail with winds from almost any direction. By the 15th century European explorers were using light, fast, highly maneuverable sailing boats called caravels to make epic voyages of discovery.

The early steamships were driven along by paddle wheels positioned on either side of the hull or, on most Mississippi steamboats, the stern (the rear of the boat). Paddle wheels were simple, but they were vulnerable to

STEAM POWER

From the late 1700s, inventors began exploring new materials and power sources. In 1783, the Marquis d'Abbans ran a steam-powered boat on the Saône River in France. Four years later Englishman John Wilkinson (1728–1808) proved an iron boat could float on the Severn River. By 1801, William Symington (1763–1831) had launched the first steamship, the *Charlotte Dundas*, in Scotland. A few years later Robert Fulton (1765–1815) set up a steamship service on the Hudson River. In 1819, the steam-powered *Savannah* crossed the Atlantic, although it relied on sails for most of the journey.

Most of a ship's rudder is under the water. It has a flat surface that can move from side to side. When moved to the side, the rudder pushes on the water flowing past it. In response, the water pushes against the boat, changing its direction.

Robert Fulton with his riverboat Clermont *in the background.*

FACTS AND FIGURES

• The first people known to row across the Atlantic were British army deserters from the island of St. Helena, who reached Brazil after 28 days at sea in July 1799.

• The first man known to sail solo across the Atlantic was the American Josiah Shackford, who crossed in 35 days in 1786.

• The first crossing of the Atlantic entirely under steam power was the paddlesteamer *Rhadamanthus*, which sailed from Plymouth, England, to Barbados in the Caribbean in 1832.

damage and inefficient, since the paddle was out of the water for much of the time. Many engineers believed the answer was a screw

A paddlesteamer arriving in the port of Savannah, Georgia. The paddle is at the stern on these boats.

LIGHTHOUSES

Before the invention of lighthouses, large fires were burned on hilltops to warn ships of treacherous rocks. The earliest known lighthouse was the Pharos of Alexandria in Egypt. Built around 280 BCE by Sosastros of Cnidus, this structure stood some 350 ft (110 m) high. Its construction was a remarkable achievement at the time and provided the model for all later lighthouses.

The 18th century saw the development of the modern lighthouse. In 1759, a lighthouse made from interlocking stone blocks was built on the Eddystone Rocks reef in England. Designed by the English engineer John Smeaton (1724–1792), this was a breakthrough in making lighthouses more resistant to damage caused by pounding waves. In the 20th century, concrete and steel took over from stone. Today, navigation technology has reduced the need for lighthouses, and those that remain are usually automated.

The shape of a ship's propeller, or screw, forces water backward as it turns. The water creates a backward thrust force that moves even giant ships forward.

propeller. This idea dated back to the ancient Greek philosopher Archimedes, who developed a device like a corkscrew in a tube to lift water from rivers. However, it was not until 1836 that Francis Smith (1808–1874) in England and John Ericsson (1803–1889) in Sweden finally came up with a successful design for a screw propeller. This proved to be a crucial breakthrough. In 1845, a tug-of-war was held between a paddle steamer and a screw-driven steamship. The screw-driven ship won easily, pulling the paddle steamer backward through the water.

Building with Steel

For most of the last half of the 19th century, the largest ships were built from iron. However, in 1900 steel replaced iron to make ships stronger and lighter, and hull plates began to be welded together, not riveted. The introduction of watertight bulkheads, which separated the hull into sections, reduced the chances of water flooding the entire hull even if a leak did develop.

Around the same time, ships began to use the steam turbine engines developed by Charles Parsons (1854–1931) toward the end of the 19th century. In a turbine engine, the flow of hot steam from the boiler spins propeller-like blades rather than driving a piston to and fro. This system converts the energy in the steam into a spinning motion very efficiently,

SCIENCE WORDS

Bulkhead: A wall that runs across a ship, dividing up the hull.
Buoyancy: Whether an object will sink or float in a liquid.
Density: A measure of how much material is packed into a substance.
Pivot: A turning point.

A ship being repaired in dry dock at sunset in Gdansk, Poland.

and turbines were powerful enough to drive huge ships. These advances in construction and engine design led to the launching of the first great ocean liners.

Parsons' turbines powered the great liners of the day, such as the ill-fated *Lusitania*. Later, liners became even bigger and faster, reaching their high point in the 1920s and 1930s, when giant ships like the *Queen Mary* and *Queen Elizabeth* made regular crossings of the Atlantic Ocean in less than a week.

Modern oil tankers, like most ships today, are powered by diesel engines, which were first introduced in 1912, or by gas turbines, first tried in 1947. Attempts with nuclear power, used for the first time in the U.S. government ship *Savannah* in 1962, have so far proved limited. In 1980, the Nippon Kokan shipyard of Japan launched a cargo ship that reduced fuel costs by assisting the engines with large rigid sails that were turned to the

SOCIETY AND INVENTIONS

Disasters at Sea

On April 14–15, 1912, the *Titanic*, then the world's largest passenger liner, sank 400 miles (640 km) off the coast of Newfoundland. The ship had been designed with a double-bottomed hull, which was divided into 16 watertight compartments. It was possible for four of these compartments to be flooded without the ship sinking. Shortly before midnight on April 14, the *Titanic* hit an iceberg, causing five of the watertight compartments to be flooded. Two hours later the ship sank. It was later found that there were lifeboats for only half the passengers, and 1,515 people died in the icy water. Disaster has also struck cargo ships. One of the most famous of these was the tanker *Exxon Valdez*. On March 24, 1989, this tanker ran aground on Bligh Reef off the coast of Alaska. Instructions to change course were not properly carried out. It is estimated that over 11 million gallons (50 million liters) of oil were released, polluting an area of 500 square miles (1,200 sq km).

The Lusitania, *a luxury liner with 2,000 people onboard, sunk off the coast of Ireland in 1915, during World War I. The ship was torpedoed by a German submarine.*

optimum wind-catching angle by computer-controlled motors. This idea has yet to be widely adopted.

Other Types of Ship

One of the limits on the speed of a conventional ship is the drag of water on its hull. Radical changes in boat design in recent years—the hydrofoil, the hovercraft, and the multihull—have attempted to solve this problem. A hydrofoil is essentially a cross between a water ski and an aircraft wing that projects on legs beneath the hull. When a

The largest ships in the world are supertankers. Some are over 1,300 ft (379 m) long—four times the length of a football field.

IRON SHIPS

There were two reasons why there was a limit on the size of ships made from wood. First, large single pieces of wood are difficult to come by. Second, wood that is unsupported over a length of 300 feet (90 m) will not be able sustain itself without breaking or bending. However, when iron hulls were devised, this was no longer such a problem. Isambard Kingdom Brunel (1806-1859) was an engineer who built many iron structures, including bridges and railroads. He built two great iron ships, the *Great Britain* (1843) and the *Great Eastern* (1858). Since then metal ships have been built to enormous sizes.

HOVERCRAFT

The hovercraft is an amphibious vehicle that rides on a cushion of air over water and land. In 1877, John Thornycroft (1843–1928) patented the idea for a hovercraft, but he had a problem—how to prevent the cushion of air escaping from under the craft. It wasn't until the 1950s that this problem was solved by Christopher Cockerell (1910–1999). He added a rubber skirt that included inward pointing jets of air.

Hovercraft consist of five main components: the skirt, the lift system, the propulsion system, the hull, and the engine. On large craft the engine is a gas turbine, and this usually powers the propulsion and lift systems. Smaller craft use diesel engines. To create enough lift, high-speed centrifugal fans suck air from intakes on top of the vehicle and force it down through the skirt and under the craft. The skirt is made of nylon and plastic, and is effectively a huge bag that fills with air. The air escapes from the bag through holes facing inward. This creates the air cushion. The bag also provides some support when the vehicle is at rest.

Modified aircraft propellers on top of the hovercraft provide propulsion. The propellers pivot for steering and there are rudders at the back. The air cushion makes steering and stopping these vehicles difficult. In case of engine failure, hovercraft hulls contain a flotation chamber that stops the craft from sinking.

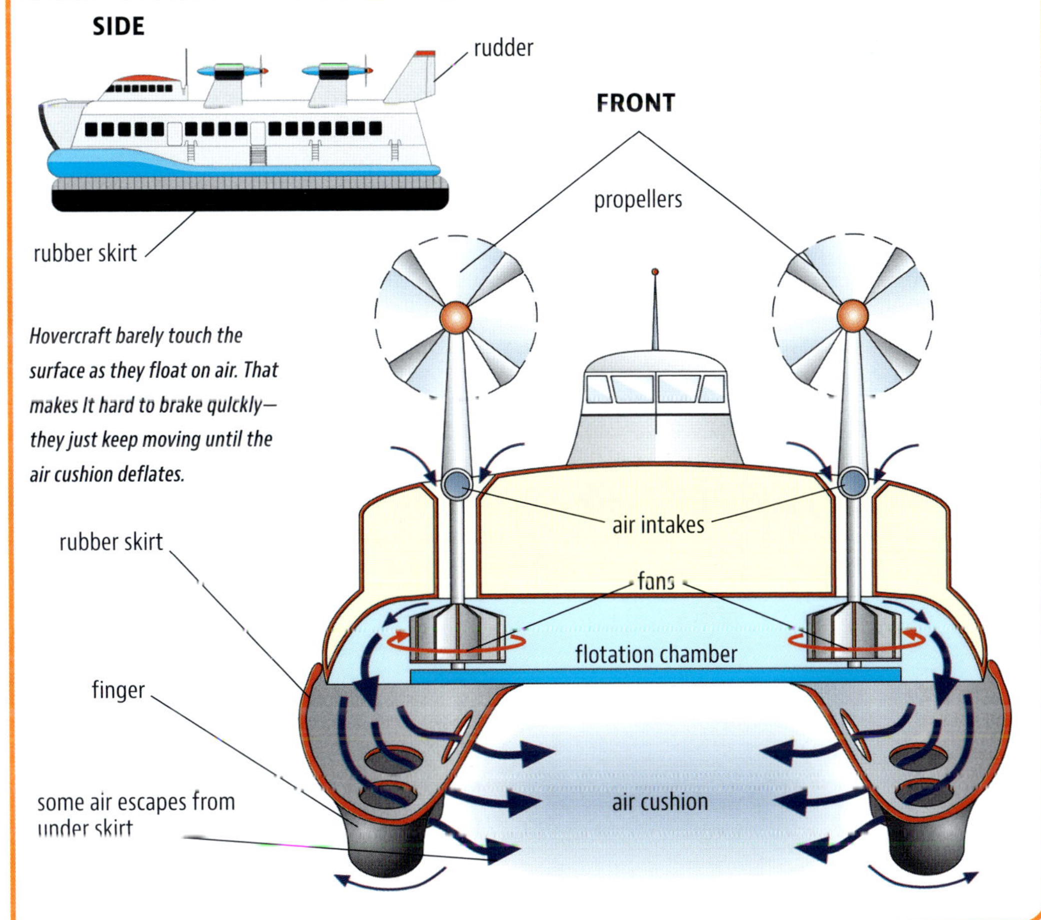

Hovercraft barely touch the surface as they float on air. That makes it hard to brake quickly— they just keep moving until the air cushion deflates.

STICK CHARTS

In the ancient world, sailors did not have satellite navigation or even compasses to show them the way. The islanders of Micronesia made accurate charts of the ocean using sticks and shells to represent the location of islands and the flow of ocean currents. Some stick charts showed the currents flowing around an island in detail. Others showed how to sail to distant islands.

One of the most intriguing future possibilities in sea travel is using solar power to drive the engines of surface craft. This solar-powered boat sailed around the world in 2012.

boat equipped with hydrofoils is moving fast enough, the hydrofoils begin to lift the hull up out of the water. Riding on a cushion of air, hovercraft can travel equally well on water and on land. The first hovercraft, built in 1959 by Sir Christopher Cockerell, was only able to carry three passengers at slow speeds and only travel over calm water or flat ground.

FACTS AND FIGURES

• The fastest crossing of the Atlantic was Tom Gentry's *Gentry Eagle* in 1989. It crossed in just over two-and-a-half days at an average speed of over 45 knots (52 mph, or 83 km/h).

• The water speed record is 300 knots (345 mph, or 556 km/h), held by the hydroplane *Spirit of Australia*.

Today, large hovercraft are used as amphibious landing craft by the military and as commercial ferries. But hovercraft failed expectations. They proved hard to steer and unreliable in rough water or on uneven land.

Future Technology

The future seems to lie with the oldest of these new ideas—the multihull. The idea dates back to the earliest boats. The Polynesians built two- or even three-hulled boats. Giving great stability and dipping only narrow hulls into the water, these catamarans (two-hulled boats) could move much faster than conventional boats. Modern sailing catamarans can reach speeds of 37 mph (59 km/h) and there are now multihulled cargo ships and car ferries. Some wave-piercing trimarans (WPT) have even been designed for military use.

THE FIRST MULTIHULLS?

Most dugout canoes are heavy and sit so low that they can only be used in calm inland waters. However, in the Pacific people attached outriggers (extra floats on the side) or added extra hulls, making log boats so light and stable that they were once used for great voyages across the ocean. The Pacific islands were colonized centuries ago in such multihulled dugouts.

BUOYANCY

The ancient Greek philosopher Archimedes realized that an object weighs less in water than it does in air because the upward push, or upthrust, of the water gives it buoyancy. When an object is immersed in water, the object sinks under its own weight, but the water pushes it back up with a force equal to the weight of water displaced (pushed out of the way) by the sinking object. The object sinks until its weight is equalled by the upthrust of the water, at which point it floats. Even boats made from iron can float because the air inside the hull makes them weigh less than water.

Modern ships display a marking on their sides called the Plimsoll line. This shows the maximum depth to which a ship may be safely loaded under different conditions. A ship's buoyancy depends on its weight but also on the density of the water it sails in. The water's density is affected by its temperature and how much salt it contains. Cold, salty water is more dense than warm, less saline water. The more dense the water, the more cargo a ship can safely carry. The Plimsoll line was devised in 1875, after British politician Samuel Plimsoll (1824–1898) saw ships carrying dangerous amounts of cargo. All ships have used Plimsoll lines since 1939.

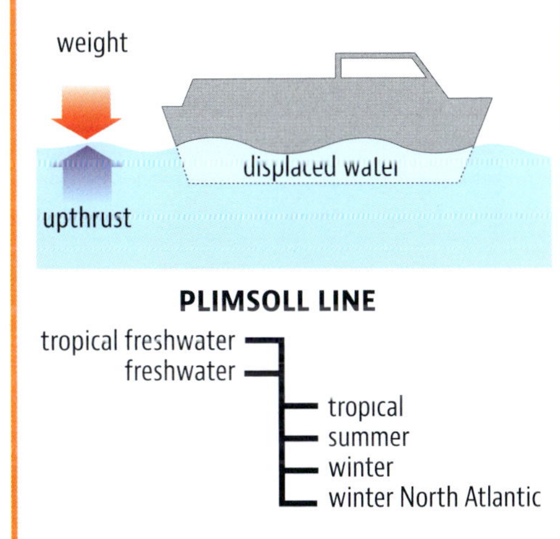

TIMELINE

3500 BCE Wheels are added to sleds to make the first carts.

3000 BCE Horses are domesticated in Central Asia and used for riding.

2500 BCE Skis invented in Scandinavia made from wood and animal fur.

1900 BCE The Mesopotamians develop the spoked wheel.

1300 BCE The Hyksos invent the horse bit, a piece of metal placed between the horse's teeth that gives the rider more control.

600 BCE Saddles are introduced by the Scythian nomads of Ukraine.

250 BCE Eratosthenes draws a map of the world that includes lines of longitude and latitude.

100 CE Stirrups are invented to help riders stay seated at full gallop.

1100s The first true sailing warship, the cog, is built in Europe.

1400 A horse-drawn carriage with a suspension system is invented in Kocs, Hungary, becoming known as the coach.

1490 A bicycle with two wheels, pedals, and a chain drive to the back wheel is designed by Leonardo da Vinci.

1493 Christopher Columbus crosses the Atlantic in a caravel, a sailing ship employing square and lateen (triangular) sails to aid maneuverability.

1550 Railroads are invented in Europe for use in mines.

1712 Thomas Newcomen develops a steam engine for removing water from mines.

1769 Nicolas-Joseph Cugnot builds the first self-propelled road vehicle, an enormous steam carriage. It is involved in the first road traffic accident when it runs out of control.

1775 John Outram invents the streetcar, also known as the "tram" in many countries.

1782 James Watt invents the double-acting steam engine, an improvement on his engines that produced rotary motion.

1801 Richard Trevithick builds a steam-powered carriage.

1812 The first practical steam railroad starts operating between Leeds and Middleton in England.

1819 Karl von Drais introduces the *Laufmaschine*, or Running Machine, the first design of bicycle.

1829 George Stephenson unveils his steam-powered locomotive, *Rocket*, designed mainly by his son Robert.

1830 The first American railroad opens in Charleston, South Carolina.

1849 The block signaling system for railroads is invented, later automated in 1867.

1860s Pierre Lallement constructs the first bicycle with pedals. The machine is called a velocipede.

1863 First underground railroad, or subway, opens in London.

1867 Nikolaus August Otto makes the first working internal-combustion engine that uses the four-stroke cycle.

1885 Gottlieb Daimler builds the first modern motorcycle. Karl Benz builds a three-wheeled car in the same year.

1888 The pneumatic tire is invented by John Boyd Dunlop.

1908 Henry Ford starts mass producing the Model T, bringing automobiles within the reach of ordinary people.

1912 The first diesel locomotive is built.

1979 The *Seawise Giant* supertanker, later renamed *Mont*, is launched. Now out of service, it was the largest vehicle every built.

2003 A maglev train becomes the fastest vehicle to travel on a railroad.

2011 Bloodhound is the first car built to travel faster than 1,000 mph (1,600 km/h).

2018 The Waymo company launches the world's first taxi service using self-driving cars in Phoenix, Arizona.

2024 *Icon of the Seas*, then the world's largest passenger ship ever built, is launched.

GLOSSARY

airfoil A curved surface that is designed to create a lift force when a fluid—such as air or water—flows around it. Wings are airfoils, and the blades of a ship's propeller have a similar shape.

ancient Greece A civilization that existed on the mainland and islands of modern-day Greece and Turkey between 2000 and 300 B.C.

benzene A highly poisonous liquid derived from petroleum, or crude oil, and used for making plastics and other medicines. Benzene was also the fuel used in the first internal-combustion engines.

buoyancy Whether an object will sink or float in a liquid.

bulkhead A wall that runs across a ship, dividing up the hull.

calliper A hinged device that can open and close in a pinch. Callipers are used in the brakes of most bicycles.

catalytic converter A device that turns many of the poisonous and dangerous gases in exhaust into safer substances. A "cat" works by pulling exhaust gases through a mesh made from special metals, which force the substances to transform into safer chemicals before they are released into the air.

centrifugal Relating to a force that seems to pull a thing outward as it rotates around an axis.

density A measure of how much material is packed into a substance.

diesel A liquid fuel that is heavier and thicker than gasoline. It burns at a higher temperature and is used to power larger engines and vehicles, such as trucks, trains and ships.

domesticated Animals or plants that have been adapted by selective breeding over many generations to live in close association with humans and perform some useful function are described as domesticated.

drag A force that opposes the motion of an object through a fluid such as air or water. Also known as air or water resistance.

electromagnet An iron core surrounded by a coil of copper wire that temporarily generates a magnetic field when an electric current flows through the wire. An electromagnet is useful because its magnetic field can be turned on and off when necessary.

exhaust The gases produced when a fuel burns. When gasoline burns most of the exhaust is carbon dioxide and water. The energy in this hot gas is harnessed to create motion in the engine before it is pushed out of the engine, and vehicle, via a tailpipe, funnel, or smokestack.

friction The force created when objects rub against each other. Friction is the force that forms a tire's grip. Low-friction surfaces, such as ice, are very slippery.

gravity A natural force that attracts two masses toward one another. The larger object pulls harder than the smaller one. Among its many effects, gravity draws objects toward Earth's surface and keeps the planets in orbit around the Sun.

Hittites A people that lived in central Turkey about 4,000 years ago. The Hittites developed many technologies, including the use of iron. They used armies of wheeled chariots.

hydraulic Operated by the movement and force of a liquid. Most hydraulic systems consist of a series of liquid-filled tubes through which a force can be transmitted.

Industrial Revolution A great change in social and economic organization brought about by the replacement of hand tools by machines and power tools and the development of large-scale industrial production methods. The Industrial Revolution started in England around 1760 and spread to the rest of Europe and the United States.

inertia The universal property of any object which makes it stay in its current state of motion—still or on the move—until a force acts to change it.

internal-combustion engine An engine that uses the hot gases formed when the fuel burns, or combusts, to create motion in the pistons or other moving parts. A gasoline engine does this, while a steam engine is an external-combustion design. The heat from the fuel is used to turn a second substance—water—into a gas, which is then used to create motion.

levitation To float in the air when an upward force is balancing out gravity.

magnetism All phenomena associated with magnets and magnetic fields. Magnetic fields are regions around magnets in which a force acts on any magnet or electric charge present.

magnets Any material capable of generating a magnetic field. *See also* magnetism.

mass The measure of material in an object.

Mesopotamia The name for the land around the Euphrates and Tigris rivers, which flow from Turkey and Syria through Iraq to the Persian Gulf. This name means "between the rivers," and the region was one the first places to have cities and civilizations.

Nazi The name of a political party or its members that, led by Adolf Hitler, ruled Germany from 1933 to 1945. The Nazis suppressed all opposition and built up

Germany's military strength. They began World War II by invading Poland.

nomads People who, instead of having a fixed home, travel from place to place in search of fresh pastures and water for themselves and their animals.

omnibus The old-fashioned name for bus. The *omni-* part of the word means "all" in Latin and so an omnibus was a public vehicle that could be used by anyone.

pantograph The conductor on the roof of an electric railroad locomotive or streetcar. The pantograph is lowered when unused, but raised to make contact with power cables running above the track, supplying the train with power.

Roman The ancient civilization that began in the Italian city of Rome around 700 B.C.

and had established a vast empire around the Mediterranean Sea by 200 A.D.

Scandinavia A region of northern Europe, usually consisting of Norway, Sweden, and Denmark. This region's history includes the famous Norse people, or Vikings.

temperature A measure of how much energy is contained within an object.

transmission An assembly of parts in a motor vehicle, including gear wheels and shafts, that transmits power from the engine to the axles and wheels.

torque The turning effect of a force, when a straight-line, up-and-down force is converted into a rotational force, such as when an engine spins wheels. A small force can produce a large torque if it is applied from a long distance, using a long lever.

A force applied closer to the pivot creates a smaller torque.

turbine A machine made from a set of blades mounted on a central shaft. A moving fluid, such as steam or air, makes the assembly rotate. Turbines are often used to drive generators.

turnpike A barrier or gate that crosses a road where a toll charge was paid before road users could continue on their way. Today, "turnpike" means any highway where a toll is charged.

FURTHER RESEARCH

Books

Cars, Trains, Ships and Planes: A Visual Encyclopedia of Every Vehicle. New York, NY: Dorling Kindersley, 2024.

The American Transportation Revolution by Aaron W. Marrs. Baltimore, MD: John Hopkins University Press, 2024.

The Future of Transportation: From Electric Cars to Jet Packs by Alicia Z. Klepeis. North Mankato, MN: Capstone Press, 2020.

Web Sites

History of automobiles
www.explainthatstuff.com/historyofcars.html

U.S. Navy Ships
www.history.navy.mil

History Timeline of the Bicycle
www.ibike.org/library/history-timeline.htm

INDEX